T0194706

WHAT IT MEANS TO BE
HUMAN
AND WHAT THAT HAS TO
DO WITH THE INEFFABLE

---

*Finding meaning and identity
in chaos itself*

---

by **M.R. HOLT**

AuthorHouse™
1663 Liberty Drive
Bloomington, IN 47403
www.authorhouse.com
Phone: 833-262-8899

Because of the dynamic nature of the Internet, any web addresses or links contained in this book may have changed since publication and may no longer be valid. The views expressed in this work are solely those of the author and do not necessarily reflect the views of the publisher, and the publisher hereby disclaims any responsibility for them.

Any people depicted in stock imagery provided by Getty Images are models, and such images are being used for illustrative purposes only. Certain stock imagery © Getty Images.

This book is printed on acid-free paper.

ISBN: 978-1-6655-6060-3 (sc)
ISBN: 978-1-6655-6061-0 (e)

Library of Congress Control Number: 2022909585

Print information available on the last page.

Published by AuthorHouse 05/23/2022

authorHOUSE®

CONTENTS

Author's note v

Introduction vii

Chapter 1 The ancient cosmos & mythos 1

Chapter 2 The head and the heart, making sense of our emotions 16

Chapter 3 The image of god 29

Chapter 4 The soul 39

Chapter 5 Good and evil 48

Chapter 6 Apocalypse 60

Chapter 7 The kingdom of heaven 69

About the author 87

Acknowledgements 89

AUTHOR'S NOTE

Dear reader, thank you for picking up this book. This literary work is an amalgamation of information contextualized in a way I have found to be quite useful. however I can assure you that there is nothing in this book that you need. My only compelling argument for my reader to continue is not one of necessity but for enjoyment itself.

I claim no profound authority, but humbly to present an alternative perspective on the quandary of human existence.

Before we begin I feel it is my responsibility to warn my reader that contrary to what you may have been told about deconstructing your faith. This undertaking is not some sexy trend but rather pursued at great personal cost. It often results in the individual loosing their religious community and corresponding identities in the process. However if you have found that those religious identities and doctoral ideologies have become more of an idol rather then a expression of a good and compassionate god, that is resembling more an egoic or dualistic persona, and one inevitably restrained by corresponding culture. then this is the book for you. Deconstructing one's faith has nothing to do with denouncing morality or desensitizing one's conscience but rather the complete opposite. Like Abraham you have decided to leave your tribe and traditions behind in pursuit of a deeper understanding of god. like Moses in opposition of oppressive and abusive authorities you too are sacrificing the stability and security of those systems in pursuit of innate human liberties, in rejection of malevolent master you are pursuing a partnership with the ineffable. in presence itself you are learning how to treat the ground as holy and whole. For those souls sincerely seeking a sanctity unrestrained by sacrament induced scarcity or spiteful sterility to sustain it, but a holiness innate to humanity, one which recognizes the profoundly prolific protestations in the compassion for the commonplace and peculiar alike. You will find what you are looking for. For those who seek to embrace these properties which preside within our Proclivity for the perverse and propensity for the profane alike. And those ready to articulate their autonomy over the erroneous agency of the inanimate. rather that be our animalistic impulses of our anatomy or aristocratic authorities of our societies. I welcome you to join me on this journey. For this is a journey undertaken with the motivation of learning how to truly love yourself and others better. I started this journey as an conservative fundamentalist seeking truth

But what I found was all that shame and guilt I had carried around for simply existing was not only unhealthy but predicated on an abusive lie. If you care more about loving your neighbors then you do about feeding your ego. Then I invite you to join me on this journey to rediscover what it means to be human.

"I would rather be compassionately curious then contemptuously correct."

INTRODUCTION

"If the path before you is clear then your probably on someone else's path"
–Carl Jung

There is this fear that you will wake up one day and won't even recognize yourself anymore, but the problem isn't that you'd have changed but rather that somewhere along the way you become comfortable with an idea and stoped rediscovering what it's means to be you! You become more concerned with retaining the security and stability of a stationary identity then you did about developing the capacity for new discoveries. But sterility dose not equate to sanctity. And any expression which is restricted, any innate hunger which is neglected will inevitably exhibit volatility, and be realized. The problem isn't the hunger but rather the systems which have starved humanity of its expression. The beautiful truth is you will never truly know yourself but shall spend your entire life rediscovering what it means to be you.

Humans are curious creatures indeed, I know this because I myself shockingly enough am one. A human that is!

And unlike any other creature on this planet (that we know of anyway) we are compelled to ask the question.

What dose it mean to be human? Or more accurately what makes you, you?

In the Tanakh we find this ancient Hebrew word "bara" or to create, tho this word is also used as to sever or cut. And as we observe, this is very much it's nature in gneisses, being utilized to separate the heavens from the earth, to distinguish between the land and the waters, and to define the light in contrast to the darkness. We see this schema repeated when god takes from the side or a part of mankind and creates eve. We get this image of the whole being divided and split into parts, the one becoming two, presumably so that the parts may be elaborated and expounded upon. And it is this action of focusing our attention and specifying types where we come to understand this term species.

tho by this definition we are in fact defined as Homo sapiens or homo erectus. But that isn't really the question we are asking when we inquire or contemplate the meaning of our existence, is it? We aren't primarily concerned with our physicality but rather with our agency and autonomy.

And we weren't the first to ask this question. People have been pondering this particular quandary since the beginning of recorded history.

CHAPTER 1

THE ANCIENT COSMOS & MYTHOS

In this age of modernity our concept of the world is primarily materialistic and predicated on our physical environment. And so when we think of creation our first assumption is in regards to the manifestation of the material. But in the ancient cosmology the manifestation of the material was secondary to the assignment of meaning and value attributed to any given object. And so it is no surprise that many mythos center our distinguishing quality as humans to the formation of complex language or the formation of the unifying narrative. Where as with only a few exceptions the more modern myths attach our identity more to our technological advancements.

One of the few exceptions to this rule would be the ancient summerians/ Mesopotamians who ground their identities in their agrarian skill. According to biblical sources the Mesopotamians AE (Babylonians) roots trace back to a man named lamech an descendent of Cain. Lamech may be most famous for his claim, and I quote….

"for I have slain a man to my wounding, and a young man to my hurt.
If Cain shall be avenged sevenfold, truly Lamech seventy and sevenfold."
Genesis 4:23-24 (webster's Bible translation)

This often eludes to not only the mark of Cain but a city of blood established by cain.

We pick back up a few chapters later where we are introduced to a man named nimrod who was a kingly figure and a mighty hunter in the land of Shinar where the Tower of Babel would have been located and what would later become Babylon. Nimrod is described as an rebel, mighty warrior or "gibor" in reference to the "gibor em" (nephelem) (Nimrod doesn't exhibit scavenger traits but predatory hunter traits) he is also accredited as founder of the Assyrian city of niniva.

In regards to the Tower of Babel.
"To the poet, the mathematicians words are babel".

These images are no stark contrast with that presented by Gilgamesh as a mighty blood thirsty warrior and that of Marduk. However much like Cain, these people were perhaps most proud of their agrarian accomplishments.

The Enuma Elish

One of the most popular and most ancient stories in antiquity is the enuma elish form the Mesopotamian civilization. This story recounts the primordial origins of creation from the volatile chaos monster Tiamat(salt water) and the vivacious apsu (fresh water). We are told Tiamat and apsu gave birth to lahmu and lahamu whom contained the universe. From lahmu and lahamu came anshar(the heavens) and kishar (the earth). It is here in the hierarchy where we find the deity anu or llu begotten by anshar. Anu created IA or enki who was to be the father of the great Marduk.

After enki and a small pantheon of other deities presided over creation, the unrestrained authorities of which began to reck havoc and chaos In the realms of Tiamat and apsu. Tho Tiamat was patient with the gods, apsu was not and thus conspired with mummu to eradicate the gods. However when the gods found out about this plot they over threw apsu and under the guise of enki they subdued apsu by way of a deep slumber and then slew apsu once and for all. After this enki and and his wife damkina ruled supreme over creation for a time. It is here where Marduk enters the scene as not only the son of the most high but as among the greatest and most powerful of the gods. Enki gifts Marduk the wind which Marduk then uses to irritate Tiamat. Tiamat is patient for a time, however after much coercion by the other god's Tiamat conceded to stopping Marduk. Tiamat raises an army of scorpion men, serpents, and even ferocious dragons,.

After much discourse among the god Marduk was elected as their champion to vanquish Tiamat and her army. Although Marduk did not do so out of the kindness of his heart but with the stipulation that he would succeed his father enki as supreme ruler over the gods.

Marduk entered into one on one combat with Tiamat and after trapping the goddess in an net of lightning he filled her lungs with fierce winds until she was close to bursting. At which time Marduk punctuated Tiamat with an arrow vanquishing the mighty Tiamat once and for all. Marduk ensnared Tiamats army of monsters and used her remains to construct the heavens and the earth. Not only did Marduk win the adoration and submissiance of the gods but also form this conquest Marduk acquired the tablets of destiny. After many grand acts of creation, Marduk erected an earthly monument to his power in the form of Babylon and as a means of alleviating the toils of the lesser gods, Marduk created man to serve the gods whims.

Another similar literation of mans origin is found in atra hasis which accounts the division of the 7 higher deities (the anunnaki) who ruled over the lower deities (igigi) which were forced to labor and toil the earth. After 40 years the igigi grew weary of their labors and rose up in rebellion. It was this uprising that prompted the creation of man as servants unto the gods.

Here we see illustrated the primordial chaos, and primeval instincts of nature expressed as both a ferocious beast and the source of life. As well as the first perceptions of humans as utilities and objects.

Egyptian myth

Among the greatest cultures in antiquity is the famed Egyptians.

In one variant of their creation stories we see many similarities to our later biblical accounts. we find the world is covered in darkness and water, (chaos) and we are told how the god RA was elevated out of the water and thus created all things by naming them. But even more profoundly is the depiction of this god RA becoming human and reigning as a pharaoh in the land of Egypt (which in both Egyptian and early Hebrew myths is synonymous with the garden of eden). But as time passes RA grows old and no longer commands the respect of his people, so after convening a secret council RA decides to begot a daughter named sekmet who will slaughter mankind. And she does. it is in this story we get a few subtle similarities to the flood narrative in genesis such as mans salvation comes by way of a beer named (sleep maker) (**reminiscent of Noah who's name means rest)** that delivers man from the wrath amidst a great flood. (This is interesting because many psychologists theorize that our dreams function as intercessors between our conscious mind and our unconscious instincts. This highlights the dreams ability to synthesize events and outcomes.)

After this event RA renamed sekmet to Hawthor and bestowed her with peace rather than violence.(symbolic of the rainbow)

After this RAs wisdom began to diminish with age. Showing that ancient wisdom must make way for new discoveries and ways of thinking. But as is often the case, the young must learn the old ways or in this myth the secret name of RA.

And thus enters isis, the wisest of the lesser gods. Isis creates a serpent to bewitch RA and coheres him into giving her his secret name thus making isis and her son Horace like god.

(A key detail is that this serpent was not made by RA or god)

This concept of wisdom, making man like god is reiterated in the tale of the book of thoth which illustrates not only the power but also the tragedy that comes with great knowing. This knowledge gave its beholder both the ability to understand all things as well as power over all things and their meanings.

Setna (the one who possessed the book of thoth) begot a son named Si Osrire. who thru listening to his fathers wisdom became even wiser then his father. This is depicted in the Nubian Sorcerer. a myth which illustrates a confrontation between foreign magicians and Egyptian wisdom.

another example of Si Osire's profound wisdom was depicted in the account of two funeral precessions where one was for a rich man and the other for a poor one. tho setna presumed the rich man was attributed glory, Si Osire recognized it was the good deeds of the poor man rather then the evil deeds of the rich that attributed glory in the after life. and further more, the poor man was more endeared to suffering and thus more apt for sympathizing with the trails of the afterlife. where as the rich man had become lazy and complacent.

The last Egyptian myth we will look at is more a kindred to job, or at. Least from the angle we will be approaching it from.

We find this ancient Egyptian story of Osiris and his brother Sit

Osiris was the oldest son of Geb the god of the earth, and Nut the goddess of the heavens or sky. Osiris was joined in union with his sister isis, and ruled Egypt teaching the people agriculture, law and ordering society, as the architect of civilization. Sit also known as Seth was Osiris's younger brother and bore characteristics of Apep (the serpent of chaos), as a storm god of chaos and destruction. Sit is warned of his destructive nature and exercised a studious disciplining of these crucial traits with the aid of Nepthys his sister and wife. That is… until he finds his older brother Osiris and his wife Nepthys entangled in an love affair. At seeing this, Sit becomes jealous and conspires to enact his fury and wrath on Osiris.

Sit throws a party and invites all the gods (with the exception of Thoth the god of wisdom and written language, and Amun-Re(RA) the sun god).

At the party Sit unveils a box and convinces all the gods to attempt to fit in the coffin, promising a special gift to the god who fits perfectly within.

Osiris is the last to attempt to fit in this box and as per its designed dimensions, he is a perfect fit for the box.

Overjoyed for winning the prize Osiris inquires as to what the reward may be.

To which Sit reply's by plunging his sword into his brother and bestowing his brother with the "honor" of being the first god to die.

As the god of destruction Sit thoroughly dismembered Osiris and scatters his body parts.

With Osiris vanquished, Sit declares himself supreme Pharaoh and is uncontested for a good wile.

isis and Nepthys swear to avenge their lover and brother Osiris. eventually they succeed in resurrecting Osiris. isis reassembled his body (Thus giving birth to the process of mummification).

And tho the resurrected Osiris's second chance at life is shortly lived, him and his beloved isis conceive a child together before Sit finds them and once again vanquishes Osiris a second time.

After this Sit continued to rule with Nepthys begrudgingly by his side.

As isis raised her and Osiris's son Horus in secret.

When Horus was grown he resolved to vanquish his uncle Sit and reclaim the throne. At declaring this resolute proclamation, Thoth imparts a bit of ambiguous wisdom about Horus and the nature of the universe.

Thoth told Horus that he was the split image or reflection of Amun-Re and subsequently a god of order and creation. But then Thoth redirects his attention to the chaotic qualities of Sit who's indiscriminate destruction had devastated the world but was of paramount importance in moderation. That in order for new life to persist the old must die to make room. Creation needs chaos. And that if Horus did succeed in vanquishing Sit, he would have to take up both crucial roles of erecting order as well as enacting chaos and death likewise.

As it would turn out Horus did succeed in vanquishing Sit.

Greek myths

The Greek and Roman myths are the ones that would have most influenced the imagery and concepts utilized by Christianity. However we will not recount the numerous myths and legends of the Greek culture but rather partake in a brief character study of the archetypal figures depicted in these stories.

In the Greek mythos we find as the main figure the god zuse who embodied this amalgamation of primal man and politician. This egotistical character who's omnipotent authority is only matched by his animalistic impulses, often even turning himself into animals and objects in pursuit of sexual partners. This correlation between our animalistic drives that perceive people as recourses and ourselves as utilities (the objectification of people) was quite profound. Zuse as the story goes was the progeny of Cronus who as a usurper of his father also feared being usurped by his offspring. And it was this fear that was also passed down to zuse. But where as Cronus was fooled and tricked into devouring a decoy instead, zuse on the other hand would not fall for such a ruse. And resorted to devouring his wife metis so as to mitigate the fortuitous usurping of his throne by porus. How ever this attempt led to the intercession of zuse's conscience. As we are told he was afflicted by server headaches that were only alleviated after festus opened zuses head, at which time a fully matured Athena (the first born daughter of metis) emerged. In fact we are even told that Athena became zuses favorite child because she came directly from him.

Zuse's childhood was presumably spent on the distant island of Crete under the guidance of gia the primeval goddess. However gia would later side with Cronus when zuse attempted to usurp the throne and free his siblings.

Zuse's entire story arch consists of the primal sexual conquests and egoic defense of his power and authority.

And thus Zuse is the epitome of patriarchal archetypal characteristics as the tribal deification of ancestors meets the political governance of the state.

Interestingly enough zuse was also a storm deity like the cainanite bal or the isrilight whyh. he claimed to control the weather, and established his dwelling on the holy mountain of Olympus.

We see many examples of giants, such as cyclops and gigantes among the Greek gods. These giants even fought many battles both against and with the Olympians and the titans.

We also see many examples of half humanoid creatures such as centaurs, Minotaurs, and harpys. These images may have depicted a more primeval form of man's evolutionary process.

Jesus did not stand alone in his archetypal role of an god in the flesh. There are many examples of this role around the same time such as Caesars and pharaohs which were embodiments or incarnations of celestial beings (deities).

Among these examples of this archetype in the ancient world was that exemplified by Hercules as the half man half god, son of zuse. Hercules was the product of an affair between the god zuse and the mortal princess Alcmene and thus incurred the wrath of a jealous Hera, who was zuses wife at the time. As a small babe Hercules foiled Heras attempts to assassinate him by playfully out whiting the serpents she sent to kill him, however Hera's vengeance was something that taunted Hercules his whole life. When Hercules fell in love with a woman named megara who bore him many sons, Hera struck Hercules with madness which caused him to destroy everything he loved, slaying his wife and children in the process.

After this tragedy Hercules sought penitence under the guidance of Apollo as a humble servant of eurystheus, who gave Hercules 12 labors or trials to complete.

This is a brilliant example of how our subconscious neurosis can lead to calamity and eventually our attempts to consciously rectify those behaviors and beliefs instilled in us as children.

Among these labors Hercules was charged with vanquishing the nemean lion which stole women from villages and then disguised itself as the women, subsequently devouring those brave warriors who tried to rescue them. Much like how our delusions of love or what we believe we ourselves need can often be the problem behind failed relationship in disguise. Another fearsome beast Hercules was charged with slaying per the 12 labors was the hydra, a 9 headed serpent beast which used its eight immortal heads to obscure its one mortal one. Aside from having 8 false heads, each immortal head severed, subsequently grew back two new ones in its place. This may have been an analogy for martyrdom.

The final labor we shall discuss here was the humbling job of cleaning a stable in which Hercules employs the aid of two rivers which did the job for him. A display of how on occasion the best solution for cleaning up a mess is to allow nature to take its corse.

Hera did not exclude her jealousy to the illegitimate children of zuse but also those who zuse pursued, one of these unlucky women was Leto.

After an affair between zuse and Leto, zuse's wife Hera became jealous and cursed Leto, prohibiting her from finding shelter in any sanctuary under the sun causing Leto to wander and be unable to give birth to her children. However eventually Leto did find an inhospitable and unstable island (later to be called Ortega) to inhabit. This volatile island became stable as a result of Letos presence and she was finally able to bare her children artmaus and apolo. This possibly served as an analogy for how a stable and secure person can positively affect an insecure and unstable environment or person thru relationship.

Hera was not the only wrathful god however.

Perhaps the best candidate for that would still belong to zuse.

When The Olympian Prometheus, famed for fooling Zeus and stealing fire from the gods on behalf of the humans is found out. Zuse convicts him to an eternity of torment at the hands of an eagle. This transgressions of Prometheus resulted in delivering mankind from its subservience to the gods.

Pandora committed a similar offense, however where as Prometheus STOLE fire, pandora REGIFTED fire to mortals and was subsequently cursed to live with humans and bare them a great number of misfortune. These may very well be among the first archetypes of the heretic. As those who took knowledge and power for themselves in order to redistribute to those without, subsequently challenging the authority of the powers that will.

It should be noted that it is from this Greco Roman culture which we derive our modern idea of omnipotent or absolute right and wrong and subsequently that idea of an all powerful, unchanging conception of god.

For it was this malevolent omnipotence which was required to progress and unify the diverse tribal demographic of the ancient world on such a scale as that which had never been seen before.

Genesis myth

The first book of the Bible offers a great deal of commentary on the prevailing narratives of the ancient world. It also offers an alternative to their assignments of value.

In this ancient scroll there are a number of literary themes. The first of which is this idea that all nations and people are related and of one lineage. And further more that all of nature, from the trees and mountains to the animals and people, came form the same source. The story starts with the shaping of the environment and the separation of the land form the waters. (The heavens and the earth). And then follows the gradual evolution of life's complexity from plants, to fish and birds until eventually we get the beasts of the field and humans. We then watch as man and woman are made conscious of themselves. Here we see the creation of culture as an attempt to shape the natural world to man's liking (the fall of man). Originally this culture was intended to help man understand their own nature and by extension that of all creation. However we watch as this authority is usurped, first by the serpent (or our animalistic impulses) and later by the Elohim and nephilim (or our deification of ideas and ideologies). thus this brings about a chaotic flood which destroys these divisions imposed by culture and reverts man back to nature. This is an act of decreation or reverting back to the original primordial state. After the flood we watch this race of humans become divided once aging as the emergence of great empires arise. And with these empires comes a dualistic conception of the cosmos and a pantheon of egoic personalities as agents of nature. And so not only is mankind divided agents itself, so is nature. It's no surprise That nature isn't often kind but violent and unforgiving. loins devour lambs. And the weak die young. And so we needed a reform in the form of civilization. However the barbaric man was even worse than nature. It is at this point in the story where we find a man named Abraham who is called by god to leave his tribe and traditions for a strange new land which god will show him. Abraham follows god and after much patience is given an heir In the form of Isaac. In the book of genesis gods blessing to the chosen ones, refers to prosperity. Or more accurately virility. (This illustrates the ideological propagation of the most well adapted and equipped genes regardless of how peculiar they may appear to the current stage of evolution and life. Favoring the inquisitive mind over the mighty physique, the sympathetic heart over the critic.) this is what Darwin referred to when he discussed "survival of the fittest" it's idealized characteristics are compassion not might which make one fit.

Another repeating theme in genesis is the one god often chooses to continue the lineage is not the first born as was the custom in that time, but rather the least of these or second borne. And so tho Isaac was the chosen one according to the Jewish tradition, he was not the first born son of Abraham. However the lineage of gods elect was descended from Isaac, the second born and not Ishmael the first born (according to the Hebrew Bible) this is not the case in the Islamic tradition which recognizes Ishmael as the elect. both sons however became the fathers of great nations. Tho these two nations would always be in conflict over this claim, thus propagating this division of humanity. We see this animosity between brothers repeat all throughout the book of genesis. From Cain And able to jacob and Esau. This motif comes to a climax in the story of Joseph and his 12 brothers, who would later become the 12 tribes of Israel. Israel is a name which means "contends or wrestles with god or el (Elohim)" and was given to Jacob after he wrestled with (an angel or Elohim).

(blessed are they who contend with the ineffable, they who entertain the unquantifiable questions of the cosmos. To believe in a world no longer divided against itself is to commune with life itself. To no longer be isolated and vilified but to see both god and ourselves staring back at us in even the most retched of beasts.)

The heritage of this family tree is interesting because it supposes that Abrahams father lived in Suméria and Ishmael is presumably the father of the Assyrians. The canaanites were presumed to be descended from Cain and yet so are the Sumerians thru nimrod and lamech.

Now originally we are told that man was made to be gods partners in creation. However when we surrendered our authority over to our animalistic impulses and our ideologies we forfeited that partnership. And that role was given to a few select people such as pharaohs and kings. However a repeating theme in genesis is the on going attempts of god trying to temper the ideology and tame the animalistic impulses. This is done thru a motif called the test. (lead me not into temptation but deliver me from evil) or essentially god spare me the test.

We see this test failed by Abraham, issac, and Jacob, and it's not until we get to Joseph that we see all nations being blessed thru this chosen people.

Before we get to Joseph whoever we have to get thru his 10 older brothers. One of these brothers was Judah who failed the test by not upholding his promise to protect his daughter-in-law. In chapter 38 we find this story often (miss quoted) to condemn certain sexual practices. However this is not only not the point of this story but also an inaccurate interpretation of its events. In this story Judah marries a Canaanite woman who gives birth to three sons. The eldest son married a woman named Tamar. However the eldest son was soon struck dead without an heir and so as was the custom in that time Tamar was wed to the second son named onan. However because onan knew that the child of Tamar would not belong to him but would rather carry on the lineage of is brother. Onan consummated the marriage but refused to procreate with her. And we are told it was for this reason that onan was struck dead too. In fear of loosing his last son to the same fate. Judah exiles Tamar and cuts her off from the tribe. This is not only shameful but essentially a death sentence at that day in time. However Tamar remains faithful and even gains reentry into the family.

In verse 34 we witness an affair between an outsider named shechem and dinah the daughter of Jacob. Because shechem wasn't of the same tribe as Dinah the union was not permitted by the customs of Israel. And so in order to enact (justice) simeon and Levi, the sons of Jacob convinced shechem and his men to become circumcised. Wile shechem and his men were still recovering from this act of submission, the two brothers slaughtered every male in the village.

(prior to the rise of Rome the animosity and fear of the unfamiliar was so great that it was even taboo to wed outside of the family or tribe and this often resulted in the marriage of cousins and even siblings being common practice, however today we have discovered the genetic defects caused by these unions. and thus thankfully this practice of inter familiar unions are taboo today. Now Ideally it is the union of a diverse assortment of genes and their subsequent traits that is proven to be the most beneficial for the offspring and society in general. However there is also a limit to this. compatibility is just as important as diversity.)

Finally we come to Joseph who is gifted with the intuitive ability to decipher dreams and there meanings. It is this ability to make the subconscious nature known, which then delivers the ancient world from a catastrophic famine.

A key detail pertaining to This "test" is it often confronts our presumptions of division and scarcity, that propagate a conception of an us verses them and a limited amount of recourses mentality. But the conclusion to genesis is, there are natural cycles of fruitfulness and famine but there is always more than enough for everyone and further more this blessing is not only plentiful but intended to bless all people and nations. Ultimately this test is a paradigm shift that involves a form of death to our preconceived notions of reality.

(Deliver us from evil)

But often what happens when we encounter something we don't know how to respectfully entertain or when we settle for something we can't truly appreciate we inevitably bring destruction on it and ourselves(evil).

Dear reader I obviously took liberties with this narration of genesis. However I believe this amalgamation dose not do the fundamental sentiments an injustice by any means but rather makes the original narrative more palatable for the reader of today to digest.

Exodus myth

This story begins with a foreign people group becoming prosperous in what was proposed to be a potential paradises. when all of a sudden the people in power decide to impose on this foreign peoples liberties. And thus this once potential paradise becomes an malevolent master and abusive authority. We watch as this utilitarian system commits infanticide to curve the foreign populations growth and degrades this people group to no more than a utility for production (slave labor). And so when we find a member of this marginalized people group adopted into an authoritative position but powerless to rectify the abusive system he profits from, he in acts justice in the only way he knows how to and is subsequently exiled into the wilderness. It is here in the wilderness where this just and compassionate character finds a tribe of likeminded people and community. And here Moses learns how to treat humanity as holy, and the ground as sacred. It is at this point in the story where Moses is called to abandon the master slave dynamic of humanity's subservience to culture (the divine idea) and is instead invited into a partnership with the ineffable rather then devotion to an idea. We are told that Moses becomes like a god to the godship of Egypt and pharaoh.(exodus 7:1-2) However this is a daunting position and so understandably Moses argues with god because he doesn't want this authority or responsibility. For you see in order to step into this role, Moses would have to deconstruct this predisposed sense of lack or innate debt posited in our humanity by cultures dualistic response to the dichotomy of nature (Us versus them). and yet this is the same reason Abraham left his tribe and traditions. For they too were governed by a pantheon of dualistic personalities as opposed to a unified and wholistic conception of god and creation. this had always been what the story was about. And so of corse god finally convinces Moses to deliver humanity from a cultures abuse of authority. However it's important to point out that this endeavor was not executed under the name or authority of any god known to man but by the authority of presence itself (I AM). This in mind, we are then told Moses liberates gods people from that system that had most certainly provided a sense of stability and security but at the expense of innate human liberties. And this transition is described as a death or act of decreation (plagues befalling Egypt). The final act of which is the death of the first born of every family in Egypt (including Israel). However there was an substitute provided to atone for this "debt" incurred by humanities compliance with culture. And this substitute is significant in its symbolic rejection of that system. Because as it was the first born who would inherit everything. it was the benefactor then who must refuse to participate in that hierarchical system that unfairly took advantage of those without power. (It should also be noted However that these systems were not constructed by these benefactors but rather inherited and thus along with them came all the harmful experiences that weren't resolved by their progenitors.) This is important because tho this idea of atonement is often described as a payment for a debt. Moreover this word atonement is more accurately described as a covering or shield from exposer to truth. (More a kindred to sunscreen in the summer or a warm jacket In the winter than it is a payment for incurred debt.) And so it is no surprise that the next thing we see is these people cross thru the reed sea (chaos water) and die to their social identities which had been inherited from their culture. This is why Egypt is depicted as being swallowed up by the waters for retaining their cultural identities. this is still demonstrated today in the practice of baptism in accordance with the Christian traditions. It's symbolic of a death to the illusory identity or covering/persona, and an rebirth or exposure to reality and deeper truth. But here's the kicker because when you've come to despise the sound of your own voice and believe you are broken, it's very hard to fix your need to be fixed or listen to what your inner voice is trying to say. It's hard to treat the ground as holy when you've spent your whole life trampling it. When you had come to believe that life is a problem that needs to be solved, you have to relearn how to simply live it as it is. And this is often exactly what we find when we strip ourselves of these personas and social identities. Without this covering we are exposed and vulnerable. And so of corse these people who were liberated have to then grieve the loss of that stability and security which those restrictive systems had provided. And that's why when we see this people encounter an very potent yet abstract expression of god at the foot of the mountain, they are hesitant when that god invites them into a partnership with them(god), and to wrestle with the ineffable as a nation of priests. Understandably they are frightened! and of corse they are! However instead of welcoming that fear as an invitation to grow, they regress and erect an idol or master (bal) to represent them. Because these old traditions and stationary identities are safe and familiar. I mean Aron even affirms this when Moses confront him. Claiming " we threw the gold In the fire and this is the shape it took" like, this is the way its always been! And so who are we to question tradition! Of corse They cling to their heritage and identities for comfort.

it is here in the story where Moses fully steps into his messianic role as intercessor, as he argues with God on Israel's behalf.

As the defining role of Messianic figure is to incur the cosmic wrath of their corresponding religions and cultures in order to intact deeper more empathetic forms of consciousness here and now (they willingly submit themselves to hell in accordance with their traditions in the pursuit of bringing about a heaven on earth). And This mirrors god's intercession in (genesis 15:9-21) when god consummated the covenant with Abraham. Because this was always the criteria. for it was that creator who posited in us a propensity for adaptation and growth and subsequently the proclivity for corruption and decay. This covenantal partnership was an assertion of our autonomy. and so with this in mind when we see god give Israel these famous 10 commandments. This is god designating the parameters for the partnership. Ultimately god is defining the relationship as a means of providing a sense of stability and security to comfort this liberated people who miss the sanctuary of their old systems. this is a new identity which god gives them to protect them or cover them from their own insecurities and unresolved inequities. However ultimately these coverings (commandments/LAW) do become idols which obscure both god's and humanities ineffable nature. this comes to a climax when Moses himself, the chosen one, strikes a rock in an display of power rather than speaking to it with compassion. And even later in the way the people of Israel cower in fear upon beholding the promised land. And refuse to enter into a world of abundance. As a result of their scarcity mindsets this people chosen to wrestle with the ineffable, subject themselves to what they knew rather then rediscover something new. Now it should be noted This land of milk and honey, like the burning bush and holy mountain is symbolic of eternal life, that is a life of rediscovery. This is not the immorality which is often presumed but rather mindful presence. This idea of heaven on earth may be better understood this day in age as equilibrium between the temporal and spacial dimensions. This is often also expressed in psychology as nature in the form of culture or culture in accordance and compliance with nature. Intuitive heart and rational mind in sync with one another.

"We often need a chaotic wilderness or space to deconstruct our restrictive identities and ideologies, in order to get to a place or land of abundance and prosperity."

Later myths

Tho they may lack a degree of antiquity in comparison to the myths we have just discussed, the imagery of later myths are a good deal more evolved and developed in many senses. For instance in the Nordic mythos we find similar images of primitive humanoids depicted as giants, elves, and even dwarves. In the Chinese mythos we get the evolved forms of heaven and earth or the ethereal and the caporal in the shape of a yin and Yang emerging from a Big Bang. According to this tradition This was cased by a deity named pangu, who in an attempt to escape the abstract state of chaos where the yin and the yang were one and indistinguishable from one another, then causes them to separate. In this particular myth we are told pangu was originally meant to hold the light and the dark together, however instead he was woken up and began to separate the two states in his attempts to escape.

We also find a copious amount of dragons in the Chinese mythos. however

In the Chinese culture the dragon is admired for its fluidity as an example for us to emulate where as in the western world we often vilify it as an chaos monster to vanquish. The anatomy of the two differs a great deal also. because Tho the Chinese rendition is anatomically I'll-equipped for flying with no wings, the Chinese dragon resembles a serpent or snake in water as it flys with a sort of graceful resonance with the heavens rather then a willful resistance to gravity.

Another story from the Chinese mythos recounts the life of a monkey king who's incessant desire to be revered or glorified gets him into trouble constantly. In one story in particular we are told that he transforms himself into an important heavenly body in order to be granted admittance into a party. The sentiment of this story very probably being that both argent pride and anxious shame are ingenious forms of signaling, that's not to say we don't honestly believe what we are communicating to ourselves and others but rather that the presumptions are predicated on inaccurate evaluations of ourselves and our environment. True humility is about recognizing that we are no higher nor lower in importance than the trees we divulge our deepest and most sincere epiphanies with.

(for my dear reader must most certainly convene councils with trees… right?)

In the Hindi mythos the primordial state and the highest state of awareness are one in the same and are expressed as Brahman. This state of reality is obscured by the divisiveness of the empirical or material world and the illusory nature of the figurative or experiential one. In this mythos we see a deity named Vishnu in a dreamless slumber a drift in the emptiness of oblivion or nothingness. Vishnu is then woken up by (the original sound) a symbol of resonance expressed as a hommmm… in this sound was an emote energy that sparked change. As a result of this awakening, a lesser deity named Brahma emerged from Vishnu.

In the Hindi pantheon Brahman is divided into a trinity of distinct deities, the creator Brahma, the preserver or positive Vishnu, and shiva the destroyer or negative. Perhaps this could better be described as the neutral, positive, and negative, forms of energy. All of which are important.

According to the Hindi mythos mankind was born of purusha. And 3/4s of our souls or essence is ethereal as we still embody in part, a resemblance or as a minute expression of purusha (god).

The Hindi traditions also teach that every experience is a teacher and we are students to life itself.

in the African mythos we get an account of the fall of man where the race of humans walk with god in the garden for many generations before taking of the tree of knowledge.

In the Native American myth (deep water) we are given insight into a world before humans. In this myth the entity called coyote warned the other animals that a change was coming and with this change a two legged race of creatures would rise up to rule over the animals.

In the Shasta myth, (the creation of the animal people). The world is depicted as a human woman who nurtures those who lived on her. Separate and apart from the earth woman was the great creator who had made her. After creating the world, the great creator shaped humans out of mud. However these people were not just humans but half human and animal hybrids. (possibly depicting a more primitive form of humanity in the process of our evolutionary process. These creatures could all communicate fluently with each other and did not need words that could be misunderstood. The humans observed and learned grate wisdom from the animals.

The Native American myths teach that nature speaks to us even today, so long as we are receptive to its prolific messages. The indigenous people of the Americas were often labeled as heathens and pagans for their worship of the earth. However they did believe in a great creator and sought communion with that great spirit thru studious admiration of nature/creation. They believe that every creature has their own individual spirit that is connected with the great spirit of the creator. The Native American tribes all acknowledge and realize that all of life is not only connected but dependent upon its relationship with all other life. Rather that life belong to a bug or the sun.

All through the later myth we find images such as ragnarock, Armageddon, and the apocalypses as catastrophic reversions to the primordial state of life. however in many cases this is less about the desolation of life and more about a reconciliation and even revival of life. It's the conscious culture and the subconscious nature being reintegrated to form a more comprehensive whole.

Perhaps this is why we love apocalyptic stories so much because they wake us up and reinvigorate our fighting spirit to survive.

It was these mythic stories in which the people of the ancient world found their identities, in these stories that they related to their lives in order to find their place in their world.

And thus these are the stories that shaped our definitions of humanity and recounted the experience of our conscious culture emerging from our unconscious nature or instincts. (The empirical attempts at quantifying the abstract expression of reality). But as we shall soon discuss, these attempts fall fatality short.

Poem

We crawled out of the abyss and have clung to to ground beneath our feet ever since. Shunned chaos and called our distain order. For civilized mans dwelling is the day and to the wild beast belongs the night.

Thus we have divided the world against itself. Light and dark, the volatile waters and the solid rock. The wise mind as master over the emotional heart.

And thus Man is ashamed of his nakedness only because he is conscious of it, but is man then ashamed of flesh or his knowledge of it.

To the conscious man hell is to be forgotten. to be nether man nor memory.

And For this reason Knowledge is blasphemy and even heresy but The unknown, unfamiliar even worse evil.

Man called the familiar common and trivial,

yet in the wasteland god spoke calling the ground holy, the common, profound the trivial, prolific.

When man inquired to what name should god belong, the reply was not EL, Bael, Ilu, or ra, but I AM. For We sought a temple and mountain for the divines dwelling but found sabbath as god's appointed abode, mere presence, The here and now, A garden where nature and culture both roam, where Tiamat was not a beast to be tamed or leviathan vanquished but chaos's herself to be spoken to with tenderness and candor.

That we should no longer aspire to stationary objects or sterile ideas of sanctity but to contend with the ineffable to abandon cosmological ark and step out into the primordial waters.

Conclusion

We have just seen a couple of archetypal roles expressed in this chapter. The first of which is the primordial state, chaos monster, source god, or Innate nature of nature itself.
The second is the conqueror, authoritative power structure, governing god, or law.
The third is the hero, savior, redeemer or son of god.

the last is the trickster, rebel, or heretic.

The chaos monster is the source of all life but is also the incarnation of death.

And so in comes the power structure to establish stability and security by vanquishing or enslaving the chaos monster. Now the power structure often gets idealized or idolized for its application of order unto chaos. But this power structure also often over steps it's authority and becomes abusive and even repressive towards change. This omnipotent power structure is not evil, it remembers what chaos was like and thus it's afraid of that inevitable chaos remerging. This is where the heretic comes in. Because the heretic seed the necessity for change and exposes the apocalypse already at hand. However the power structure rejects the heretic and denies chaos, inadvertently submitting itself to a form of atrophy and decline. It is only the hero who has the discipline to tolerate the omnipotent power structure and the compassion to understand the heretic, who is then equipped to intermediate between the two and ultimately bring about revival and redemption. The heretic knows only truth and the power structure knows only law, but it requires the saviors grace to reconcile the two. Ultimately what's happening is the power structure is trying to deliver mankind from the destructive nature of chaos. And the heretic is the one trying to deliver mankind to the rejuvenating nature of change. The chaos is both the destructive monster and the utopian destination. The problem is both the power structure and the heretic only see one of its roles. in the ancient Canaanite literature these archetypal images constructed a divine council which presided over the cosmos as a unification of dichotomous voices and agents. This personification of the cosmos mirrors our own internal state. this divine council of our psyche can be expressed as follows. the ego(protector), the skeptic(heretic), compassionate heart(hero), and the instinct/intuition(source/nature) however like the cosmic variant this is not a dualistic pantheon but a dichotomous conglomerate of states became unified. And each and every one of theses voices have a valid seat at the table. This is ultimately the question our ancient myths attempt to resolve. However as cultures evolve, the desired traits do too. And so we find these myths become outdated and in some cases counter intuitive.

CHAPTER 2

THE HEAD AND THE HEART, MAKING SENSE OF OUR EMOTIONS

The abstract expression or artistic image is much closer to reality than the empirical attempt to quantify it.

We all have a very different vernacular for essentially the same phenomena. For instance this word Chaos can be expressed in a number of ways ranging from destruction to ambiguity in its mythic sense. The word itself is not a constant but rather dependent upon its context for deriving meaning. Much like a woman can be both a mother and an wife and even a daughter or sister depending upon her relationship with the people around her and yet she is none of these things in actuality.

Another important thing to remember is rather it pertains to philosophy or physiology, evolution is about baby steps in the wright direction. It's small changes that produces the most precise and effective results. And rather it's our anatomy or our doctrine. Progress is slow.

Rather it is the philosophical quandaries or physiological adaptation the process of evolution is just that, a process. and is about baby steps in the right direction. This idea can be equated to being in rhythm with the music, rather we are playing an instrument or dancing, timing and tempo is important.

This is the the primary role expressed by the claim that man are partners in creation with god. The technological rational culture of man attempting to be intuitively in tune with the progression of nature. This is what is depicted in the ancient scriptures. rather you believe the events in the Bible are historically accurate or metaphorical embellishments. the Bible is depicted with mythic imagery and language. And this can be observed in the progression of slavery and racism. because in the ancient world people owned other people in much the same way people today own animals as pets and livestock. They looked at humans that they deemed (lesser) with much the same distain as we do our pets. This doesn't make such prejudice acceptable but rather functions as a reminder that less evolved society's of the past can't be judged by the standards of today. And truthfully people of the future may very well look back to social standards and norms of today as barbaric and inhumane. there's a difference between believing the rain is good or bad, and refusing to believe in gravity after you've seen it in action.

Evidence discredits ignorance. In much the same way we can't know something until we know it. Nether can we refuse to acknowledge the evidence after we've been privy to it.

The famed psychologist Carl Gustav Jung spent the majority of his life studying the remnants of primitive man hidden in modern man's psyche. More accurately attempting to unlock the prolific messages in our subconscious and dreams. Jung expounds upon the significance of abstract and archaic imagery as an intercessory of our innate nature, instincts, and intuition in our empirical rationale. And because our natural instincts are far more adapt and attuned thru years of evolution, our subconscious states AE the abstract or artistic expression is far more comprehensive and accurate (that is closer to reality) then our attempts at empiricism. It is for this reason that the facts and figures are in many ways illusory and the narrative/ abstract expression more a kindred to reality.

How ever the abstract image is far to vast for us to be able to quantify and so for this we need an empirical format or formula.

This by no means is intended to denote or impoverish the importance of our rational critical thinking skills. But rather to acknowledge the profound wisdom and information present in the abstract expression and our intuitive nature AE our emotions. Because our emotions, tho most certainly vulnerable to mutability and fallible, are also far more equip for picking up on discrepancies in our environment and communicating our needs to us then the attempts of our rational mind to repress or refute those impulses.

Even at this day in age, the ability to listen to our instincts and understand what our emotions are telling us is of the utmost importance.

As a few common examples of this relationship between our primitive instincts and our modern rational we find the iconic images of The heart and the mind, sense and sensibility, the body and the spirit, heaven and earth.

And one of the more well known narratives that depict these states in mankind is that image of the garden. That place where god walked with man, where heaven met earth, where nature and civilization coexisted in equilibrium.

And in this narrative of the garden as Jung describes, we see god create or more accurately separate and distinguish between the light and the darkness, for the day is mans abode and the night where the the wild beast dwell. The light is our conscious state and the darkness our unconscious instinct. In the day man knows but at night man dreams.

Now ultimately the goal for mankind is to merge these two plaines. To make the dream a reality, to bring heaven to earth, to understand the emotional hart thru the rational mind, for nature and culture, civilization and ecosystem to coexist in harmony.

All matter is energy and energy maters

Thru the view of modernity our conception of the world is materialistic (that is comprised of mater) and so it makes sense why we would view creation then as the manifestation of the material AE the Big Bang, or Intelligent design, etc.... But in the ancient world creation had far more to do with assigning meaning and value to that which was previously unknown. And so understandably creation would have taken place thru the act of speaking, or the construction of complex languages and the formation of unifying narratives that allowed ancient humans to communicate and congregate on a much larger scale then our less cognitive predecessors.

And because of this definition of creation pertaining primarily to such assignments of meaning and value it would make sense why figures like paternal ancestors and authoritative governing parties like kings and pharaohs were deified for there discoveries, not to mention as a coping mechanism for dealing with grief and our own mortality as observed in the epic of Gilgamesh.

But this is also why creation is often depicted as an epic battle between those inciting order and the primeval chaos monster. This attempt to tame not only the wild beasts in the dark wilderness, but also that impulsive nature within ourselves.

Or as it is described in Genesis (man shall rule over the beasts less the beasts should rule over you).

And so we see these massive strides to shine a light into the darkness, to vanquish the leviathan and Tiamat, to explain and understand the mysterious nature of nature. These are all images of our rational mind wrestling both figuratively and literally with the nature of our environment. And tho these strides are admirable, sometimes chaos wins.

The chaotic flood waters

Nearly every human culture has a flood narrative. This story of a catastrophic flood consuming the earth. Now interestingly enough most of the more ancient creation narratives depict the world as being covered entirely by water. And wether this has ties all the way back to our primitive form as aquatic creatures passed down thru our instincts to us now, or if this instead is just an common image of chaos, primordial life. One can only speculate, however the later is a common consensus by scholars today. For the ancient world, the land like the day was man's dwelling and the water like the night was where the wild beast or monsters dwelt. Water was synonymous with chaos and even the heavens (the abyss). some even believed specters, sirens, and wraiths, hovered over the waters. Regardless water was an image used to describe the world devoid of order and structure.

And so this flood and even the plagues of Egypt were images of decreation. The world reverting back to its primordial state.

A key point to acknowledge here, is much like our unconscious instinctual state, the waters are where life came from and even still a critical source of life for all life forms on earth. The problem being however, that these primordial states are far too raw or pure for us to tolerate. Much like the abstract expression is far too robust and comprehensive for our empirical systems to quantify or 100% pure oxygen is ultimately fatal to humans. So to life in its purest form is very potentially problematic.

The point here is that chaos in its ancient sense is actually life, not death, or perhaps more accurately it is death by way of life.

It's a form of life that is far too pure or raw for us to tolerate.

And so if chaos, the primordial original state of life, and the unknown is depicted as water. It makes sense then that our images of order would be dry land, as our firm foundation to build our empires on. How ever, even there the storms come and the floods rise. Even tho our cities are safe and offer stability, they are none the less small and stationary. So we need a cosmological vessel in which we can traverse and navigate the endless expanse of the chaos waters. And so for this we construct an ARK or boat. A vessel which we can use to navigate and traverse our cosmological terrain. And so this image of a boat makes for an excellent metaphor, in reality we need that which only culture can give us. A doctrine or ideology. A cosmological map of our terrain. And tho we often think of these tools as being separate from nature. The truth remains that human nature is by extension nature. And the primitive forms of these maps are still observed in mankind today. Tho instead of referring to them as refined tools like doctrine,,ideology or culture we often refer to this instinctual primitive remnant as our ego. It is the stories with which we identify with, the way we describe our role and place in our environment.

But just because this ego is seated in our subconscious and rooted in our rational attempts to quantify this world and our place in it, that doesn't mean that it is a trust worthy source. Because far to often this ego is rooted in inaccurate or incomplete stories. and then reinforced by our subconscious in very negative ways. More accurately this may be described as our unconscious trying to attach itself to the conscious, rather then the conscious mind trying to understand our subconscious instincts.

It is an exultation of the simplified map over that of the far more expansive expression or terrain.

a prime example of the flaw in this form of attachment can be observed in the way we are prone to getting defensive whenever an idea or set of beliefs of ours is challenged or refuted. Ultimately the reason for this is on a subconscious level, we've attached our identity to these ideas or beliefs. We've attempted to reduce the vast terrain to a simplified maps of it. And further more we've made our intuitive nature which is far more comprehensive, subservient to our cultures attempts to quantify those qualities in ourselves and in the environment.

We've reduced and repressed our emotions and instincts to exalt our rational mind and culture to a position of authority.

Now don't get me wrong, we should not let our emotions control us. But we should also learn to recognize what they are telling us.

"People want to be fixed, but what they really need is to be loved as they are (utmost by themselves)

Don't settle for this idea of perfection or holiness, but rather cultivate the compassion and capacity for wholeness.

The point is not to silence your emotions, but to appreciate what they are telling you."

Cells and signals

There is a wealth of information stored in our genes. From the experience of biter tastes evoked when we eat vegetables that had evolved defensive mechanisms like toxins to defend against bugs and herbivores. to attraction to physical features that indicate fertility in potential mates. And tho these genes do not control our behavior, they do incentives certain traits that have proven to be beneficial in the past at propagating those particular genes. And so ultimately things like a sweet tooth that incentivized the pursuit and consumption of foods and resources high in sugar and fats, which would have provided a hardy source of protein and vitamins especially in conditions where resources were scarce or scanty, actually served a vital purpose in our developmental process.

But there is far more than just nutritional data here.

Among this data there are elaborate subconscious processes and even incredibly effective forms of communicating with others. And every one of these genes were chosen thru trail and error in the hopes of propagating those genes And by extension ensuring our survival.

so It makes perfect sense how a fear of the unknown and identifying with the known should be extremely beneficial for our survival. However it is that same unfamiliarity which poses us a threat that also presents us with abundant potential for adaptation.

There have been countless studies done observing apes coming into contact with serpents or snakes. And in each case the apes show great interest and fascination with the serpent. This curiosity is only matched by the caution that attends it. And so much like in the garden narrative, the serpent is cleaver, it's fascinating because it doesn't look like us and it doesn't interact with the environment like us. And it is that very fact that both makes it useful and dangerous. many argue this is where we first see our ego emerge as a rejection for that which is different, or as an attempt to protect us from the potential dangers the unknown or that which we don't understand poses.

And so how do we distinguish between potential danger and that which is useful, how do we differentiate between the parlous and the playful.

Another phenomenon observed by scientists thru a multitude of studies is that of laughter.

Laughter is one of the most primitive and subsequently most innate forms of communication. In fact many evolutionary psychologists theorize thru conjunction with countless studies observing the behavior of small children prior to the construction of complex linguistic skills as well as in many primates, that laughter is most likely our psyches way of communicating to both ourselves and others the lack of danger in potentially threatening situations. More over it's how we distinguish between playful and parlous circumstances. And this is most likely the reason why most jokes have an initial shock value. And also why laughter often seems to defuse stressful or tense situations as well as alleviate anxiety. It is our instinctual way of discerning rather to approach a given circumstance with caution or with curiosity. And for this reason it is an indispensable tool.

Another primitive tool we have for communicating is that of eye contact. You may have heard it said that the eyes are the windows to our soul. And tho that may not necessarily be true, much like laughter this form of communication is so ingrained in our psyche that it is very hard to effectively fake or synthesis ingeniously. That is to say that in most cases our subconscious is so finely attuned to the authenticity of these messages that it immediately picks up on any discrepancies or attempts to mislead.

This may in fact be why maniacal laughter is so disconcerting. We have grown to trust people and creatures who can initiate and retain eye contact thru the duration of conversation. One reason why this tool may be so effective or trustworthy is due to the visibility of the whites of our eyes and how that then allows us to subconsciously communicate a degree of focus or interest, that is a degree of transparency as to where that attention is directed, and what the nature of the intentions in question (rather parlous or playful, exciting or frightening) are.

These are just a few of our innate tools we use for signaling.

Side note! tho we may often be unreceptive to these prolific messages, as a result of our cultures attempts to suppress and repress the necessity of our physiological needs. The truth of the mater remains, we need these simple interactions such as deep meaningful conversation and intimate hugs just as much as we do food, shelter, and rest. In fact I truly believe tools such as dancing and laughing are indispensable when it comes to regulating our nerves systems. Further more an important distinction between feelings of tiredness and exhaustion should be noted. Because one is pertaining to a lack of stimuli and the other an over abundance of it. Energy is energy! Rather it is the presumptively negative energy of stress and anxiety or the supposed positive energy of excitement and joy. And thus the most effective methods of alleviating a diss rest in our nerves consists of simply expending and exhausting that energy, stress, and anxiety thru exercise, dancing, laughing, crying, jogging, or any other form of activity which provides these energies an outlet to dissipate. And so it's no surprise that the tactics of repression and regression tend to make the issues worse rather then better.

Sacrifice

Another form of signaling is our ability to expend vital recourses frivolously. This is especially useful in dominance hierarchies where one's ability to sacrifice crucial recourses or expend vital energy communicates a plentiful supply of energy and resources on the individuals part. This is also a useful tool in asserting allegiance to a cause or community in the form of a sacrifice which not only communicates to the tribe, ones allegiance but also solidifies the convictions of the individual to that cause. For we are far more likely to commit to something we've already devoted time and material to then to cut our losses and find a better alternative. The reason for this is we feel as tho our recourses were wasted or lost if we abandon a faulty cause. And tho this often results in "throwing pearls to swine" if you will, it is much easier for us to chalk it up to a test of faith then to admit that we were wrong all along, which would imply that we could be wrong about other things as well.

For instance The incans believed that the red sun was fueled by human blood and so for this reason they were constantly sacrificing or waring with other tribes so as to bring about the next day. And tho today we think of this as ridiculous, for them it was an assertion of their autonomy and power over fate. Because in their world the rising of the sun and bringing about of the new day was directly dependent upon their actions.

Now tho this takes a much different form today then it did in ancient civilizations. We still signal thru lavash lifestyles, luxurious cars and houses, and ostentatious donations to causes among the rich.

And tho these forms of signaling are primarily subconscious, we can benefit greatly from recognizing their meanings and communicating those prolific messages into our conscious awareness.

there have been studies conducted where scientists put two different rats in two identical wheels. the only difference being that one rat controls it's wheel where as the second rats wheel is controlled by the firsts movements. what they found is the rat that chose to run had all the defining markers of someone who exercises, where as the second rat that had been forced to run exhibited traits of someone with severe anxiety or stress. The adversity they experienced was identical but the mentality of the one who exercised its autonomy was much improved to that of the one which had suffered that adversity. We see something similar in the ancient Canaanite traditions where a transition of deity worship takes place, passing form the ancient one who had inherited agency to that of the master who had exercised their authority in obtaining power. We see this described later in the exodus myth with an invitation to abandon that master servant dynamic for a partnership with god. Often this form of sacrifice functions as a vessel for asserting the autonomy of the individual over fate. The point being things do happen to us as a result of other peoples choices, but our reaction to those events are far more consequential then the events themselves. Ultimately fate is just the result of our habits and beliefs producing their subsequent outcomes. And the assertion of our ability to change those trajectories is of paramount importance. To chose not to be victimized by our own subscription to a set of beliefs is how we exercise our authority over this outdated form of cosmic agency. Subconsciously this mentality is most commonly communicated thru confidence as the uncertainty of any given circumstance solidifies one's autonomy rather then threatens it.

Fear

There is far more to our genetic coding than commutative tools. There are also impulses like fight or flight impulses and attraction/reproductive motivations.

Now before we continue let's talk about fear. Because fear is often associated with very negative connotations, but not only is fear crucial for our survival it's also absolutely beautiful. Yes beautiful!

Because fear is what you feel when your in love with someone, those butterflies in your stomach, that racing heart, that take your breath away feeling. Fear is what you feel when you stand at the precipice of the Grand Canyon or when you look out over the ocean waves crash violently against the rocks. And fear is the reason you jump out of an airplane at 14,000 feet. Fear is where awe comes from, in fact awe is just a different response to the same core emotion. Fear tells us to pay attention because what we are witnessing is much much bigger than us, and it presents us with the opportunity to grow and learn something. To be prevê to a world much bigger than our own. In the Bible this idea is referred to as the fear of god.

Now like I said fear is what we feel when we are in love and that is true. But there are also contributing factors that motivate us to pursue that feeling or person which that emotion is associated with. And that quite simply is attraction, and that includes but is not limited to physical,intellectual, and even emotional characteristics.

Now in all three cases our attraction is based primarily on traits we observe and propose to be positive. But our attraction is not entirely up to us, that is we don't always get a choice in who we do or do not like. And a great deal of this has to do with our genes or perhaps more accurately the information in our genes emphasizing traits that have been proven to propagate certain genes in return. Traits that signify fertility and virility, traits like speaking skills and physical fitness, problem solving skills and ingenuity, as well as emotional aptitude or the ability to even nullify emotions and impulses in one's self. And the quality of these traits have a lot more to do with our development and who our progenitors were then it dose our own choice in the mater.

However this dose not mean we have no choice in who we will end up with. As is often the case, as we grow we develop and even disregarded different tastes in personalities, foods, and even activities and habitats. People change and we are no exception. In fact that is the very premise on which our genes are certain of. Because the very reason those genes or their corresponding traits survived was their ability to adapt and grow.

So just because this is the way are bodies work, doesn't mean that we just are who we are and have no say in the matter. Quite the contrary in fact.

"Neurons that fire together wire together, neurons out of sync fail to link".
(Donald Hebb)

Tho our genes do most certainly hold a wealth of knowledge, it is our conscious use of that information that shapes who we are.

The muscles we exercise get stronger and those we neglect atrophy. As we grow to value certain traits in ourselves we will also adapt an attraction to those same traits in others.

Burning bushes

The point is it's hardly ever as simple as it appears.

From the fact that individual fish have more in common with other land roaming species then They do with other fish, to the fact that nearly 50% of our genome is indistinguishable from all other life on earth. At the end of the day all of our empirical definitions and titles are just attempts to quantify the abstract. ultimately they are just descriptions of reality and not reality itself. Or as the Dow put it "we often mistake the finger, for the moon it's pointing at". And so tho Ecclesiastes put it best when it said

"there is nothing new under the sun"

that dose not change the fact that we still have more then enough to discover here and now.

Now why is this important then. Well we often have a proclivity for invalidating our own experiences in light of those we see as authoritative in compassion. We hear someone call something sacred or holy and then deduce that the ground on which we stand must then be less than holy.

So what then makes something holy or sacred? The short answer, scarcity. Our minds develop a fascination for things that are different or peculiar. And we see this in culture in the way certain body parts are coveted because they are private. And yet at the end of the day it's all just skin. The same skin as everywhere els on your body. Something I noticed two years into the covid pandemic, was how attractive the lower half of the face had become. And I realized it was solely because we had begun covering or concealing it from public view. And so with this came the connotation of intimacy associated with that portion of the face below our eyes. And so often what culture will do, is attempt to retain that sanctity thru cultivating an artificial form of scarcity. however the problem with this is not only do we then question the validity of sanctity in the common place, but we also inadvertently cut the "sacred space" off from the rest of the world. And so what often ends up happening with traditions or dogmas that are considered sacred, is they often end up becoming sterile. In our attempts to immortalize them we ultimately kill them because they are not allowed to breath, to grow and to change.

Take for instance the image of the undying. Rather it is the rotting corps of a zombie (man returning to an unconscious state), or the unquenchable thirst of the vampire. Tho yes they are spared from death, they are also subsequently completely devoid of life in the process. It would appear rather in the physical body or in the figurative meaning associated with our psyche, our proclivity for corruption and decay, and our capacity for adaptation and growth are in fact one in the same. Life by very nature is death.

In the Jewish tradition there was this event called the year of jubilee that would occur every 7 years. In this tradition every Jewish person who had sold themselves in to slavery to pay off a debt would go free. Not only that but also all debts would be forgiven, and all land would return to its original family or people. They would also let the land lay fallow for one whole year. No crops would be sowed, no vegetation reaped. And tho we don't observe this tradition today, there is some significance in the fact that also within the period of a 7 year span our bodies will completely decay and regenerate a whole new set off cells. That is to say, every 7 years our bodies will have died and be completely reborn as a whole new set of cells. All this to assert that we are not stationary objects but dynamic beings that change and grow.

If that then is the ineffable quality of humanity. What then dose it mean to be human? This question is by no means a new one, but one we've been attempting to answer for many a millennia.

And nearly every civilization in the worlds history has a unique answer for this question. For example the ancient Sumerians and Mesopotamians attached their identities primarily to agrarian skills. Where as the Egyptians and Assyrian tribes found their answer in the formation of the unifying narrative or the construction of complex language that then allowed mankind to congregate on a much larger scale and work towards a common goal.

And tho there are variations all throughout history, these two conclusions seam to be at the root of all of them from the mesoamerican belief that man was created from maze (agrarian) to the Chinese account of Nuwa (nugua) the mother goddess who after seeing her reflection in water, scratched out dirt and mixing it with water creating humans from clay to keep her company (unifying narrative or more accurately communal). The two options we are left with then appear to be our utility in a technological progression to survive. Or as a member in a community or collective. But can our identity really be reduced to its utility or communal properties. Are people a means to an end. Property to be objectified or pre destined characters in a story completely devoid of our autonomy?

I think not! No it is my firm belief that we are in part both, but also so much more.

I don't think mankind was created to serve the systems put in place to serve man. Rather that be a story we tell ourselves or were told growing up, or simply tools for discovering the most efficient way of performing a task. People are not property nor are we characters here to simply fulfill a role. And so we return to the ineffable. In the hopes we might find our humanity.

A quick tangent on style. Because we all have a style. But that style is more then just the cloths we wear or the music we listen to. Our style is how we see the world, and how we see ourselves in it. We drees for the roles we want. Or more accurately we fill the role we believe we fit. The way we express ourselves thru the cloths we wear and the music we listen to is a direct result of where we think we belong in the world. We look like our environment or we perceive our environment with internal implications.

frued believed all human drives were universal AE all our actions were motivated by sex or power. but Jung recognized that the same archetypal expressions and impulses could have a broad array of causes and meanings in different people. That each case is unique. Ultimately all archetypal imagery is fraudulent or illusory due to the fact that it is by no means the thing it's self but rather an attempt to allude to an experience or phenomena. And so this shame or feeling of being a fraud because we've taken on or been given identities that are not only inauthentic and ingenious but also in many cases a more kindred to an out face lie. But it's not until we find a sense of resonance with a song, message, scent, or encounter, that we actually question these archetypes and our association with them.

the closer a creature is to nature the more in tune or in sync they are with the changing weather of their environment. This is most probably due to the fact that Life is cyclical by nature. for nature this functions as seasons like winter and spring, death and rebirth. For people these cycles can vary from our emotions to stages in life and even simply habits like eating and defecting, sleeping and waking up. Life is not static or stationary and nether were we meant to be. What spouts up stream flows down stream. Seeds sown up wind are carried down wind. Volatile climates and Inconvenient weather cultivates beauty and life. As is often the case the most passionate people are also the most temperamental. They are immeasurably compassionate and affectionate but at the same time they are often quick to out bursts of anger and depression. This chaotic relationship is no coincidence but the very nature of life. The point is not to calm the storm but to find pace and listen to the prolific messages of the storm. Not to emulate but resonate with the chaotic nature of life. It's all reciprocal. And so nether abstinence or excess is particularly favorable to this crucial relationship or nature of life. For every feast there is a famine of equal significance. You may say the logical conclusion then is to be neutral but that's not how life functions. We find life in the extremities. It's where those extremities meet that we find resonance. Rather it's a hot fire on a cold nite or a cool shade tree on an hot summer day. They work in tandem. The elate joy of falling in love in-riches the sorrow of a broken heart and vise versa. We were made for the full spectrum of life's expressions.

there are not negative emotions, there are only negative reactions to uncomfortable emotions.

even feeling like anger aren't inherently bad, there can be justice in anger, anger can be compassionate and kind. Anger doesn't have to be rooted in fear, anger can be brave. It doesn't have to result in hatred, it can often be the means for healing. Which brings us to our next point. Just because it hurts doesn't mean it's bad! Something can be painful, offensive, and uncomfortable and also be healing and good. Growth often hurts.

just because there is something more abstract behind the archaic beliefs doesn't then diss prove the belief itself but rather advocates for something far more profound. the scientific explanation for superstitious behavior is more prolific then discrediting.

Tangent

evolutionarily the distinguishing characteristics that separated humans from the other animals was our ability to reallocate important recourses from our digestive systems and to our big thinking mind. This mental ability that allowed us to be conscious and judge between favorable and unfavorable circumstances more critically, is most certainly the sinful gene we inherited from our ancestors. As It is this particular trait which is singled out In the mythic depictions of man's fall from grace. A part from supposedly separating man form god, this reallocation of recourses also prompted our need to cook meat before we consume it. However it should be noted that it is this very same (sinful gene) which is also the root of mankind's most distinguishing quality, our ability to love and have compassion for other creatures.

CHAPTER 3

THE IMAGE OF GOD

Another distinguishing characteristics of mankind is as the barer of gods image. An attribute perhaps most commonly allocated to humanity thru passages like genesis 1:27. However this is a fairly modern understanding of this claim because in the ancient world the divine image was only attributed to those in an exalted position of authority such as kings and pharaohs. And similarly this is where we get this idea of idols. Because the image of god's were a idea to aspire to, not something innate to all humans.

In the ancient world these clams about divine entities or gods were not concerned with ether validating or disproving one god or the other. But about aligning oneself with an assignment of meaning and value which served ones values. It was about asserting one's perception of life's nature. Is god just or merciful, does god care about our goals and values or there's? Who's side is god on. Ultimately is god big enough for both. Because that's the cool thing. We all have very different purposes and goals that motivate us. Some of us are here to conserve and preserve, and others are here to progress and expand. The question is "is god big enough" for both those in authority and those marginalized. Who's god offers both justice for those abused and oppressed and mercy and compassion for those who were misled. The question was not about invalidating a gods authority or power but rather about seeking a moral standard that works for all of humanity. Tho as you've probably realized by know. These standards are often handed down to us by our parents or the culture of our given society. So it obviously makes since why many of these ancient conceptions of god are just projections of our our parental figures and progenitors. And so they can often be very bias. This is why we must make distinctions between these dualistic projections and the wholistic nature of creation. We need a god who is not restricted by culture or class.

With this in mind an important question may arise quite simply "who or what is god?

The term god as observed in many other ancient cultures was a way of immortalizing those in a omnipotent position of authority or those ancestors and progenitors who had made significant contributions to the structure of a society or culture. It is for this reason that Many evolutionary psychologists theorize that this conception of god originated as a coping mechanism for not only dealing with the loss of loved ones but also as a means of retaining their assignments of meaning and value for any given culture. And it may have even evolved into a means for the psyche to assign agency to those ambiguous messages in nature and our minds incessant need to attribute meaning to things.

Despite this apparent projection of deification on ancestral figures of authority. This hypothesis dose not invalidate the existence of deities or god but rather highlights a phenomenon observed in virtually every culture in human history.

Evolution has instilled in the human psyche an innate bias towards assigning agency to phenomena in our environment. And this is most likely a mutation of the mechanism intended for computing causal scenarios and hypothesizing the effects of circumstances in our environment. Not to mention extrapolating the repercussions of our own actions or reactions. And so it is no wonder this mechanism would produce a surplus of seemingly prolific information. This form of intuitive cognition is innate to humanity and most certainly serves a vital role in propagating our existence. However this dose not undermine or demean the importance of cultivating an reflective cognizance which expels those prejudices and biases.

since the dawn of time humanity has stared up at the stars and deep within our being we have felt the universe staring back at us.

In perhaps it's most original state the answer is still quite ambiguous. From the Mesopotamian jinn (a word which we now derive genie and further more genius from) to the biblical depiction of Elohim. a title bestowed on both the ancient one el or Adini and those members of the divine council more commonly know by the name angel (more simply Greek for messenger). These images quite simply translate to a muse or transcendently inspirational force.

In fact among these deities would be included the opposer or satan, who was seated in the divine council (job) and may have even been given a portion of creation to rule (Ezekiel 28). But more then likely fills the role as gods executioner and would not have been depicted with quite the same distain as is often attributed to what has come to refer to a singular deity as opposed to its original usage as a description on an archetype.

However in the Hebrew Bible anyways, great strides are made to clear up a lot of this ambiguity. And we watch this idea of god evolve from a pantheon of divine figures of which El or (the ancient one) is head, but who later relinquishes his throne to whyw or the lord (bal) in psalm 82. We whiteness a paradigm shift of which we find insurmountable evidence of, from ugaritic and Akkadian iconography and artifacts dating to around the mid to late Bronze Age 18-14,000 BCE all the way up to around the early Iron Age 8,000 BCE. Further more we can observe this transformation take place in the narrative of the biblical accounts. as Abraham is called by god to leave his tribe and traditions behind and sojourn to a new land. How Abraham is required by el to sacrifice his son Isaac but at the last minute is stoped by whyw or the lord who offers a substitute in the form of a ram. (Abrahams god didn't require or condone human sacrifices).

Or how god calls mosha to deliver his people from bondage in Egypt, (which at one time was a potential Eden and divine form of order that god had used to deliver the ancient world from an catastrophic famine.) But that had become a supreme form of order that had been manipulated to abuse and oppress the people it was meant to serve.

And so god allowed chaos to befall Egypt but delivered Israel thru the chaos waters and into the wilderness where they could deconstruct these practices and beliefs before entering into a land of promise, so as to not propagate the same injustice as their captors. Because before Egypt abused the Hebrew slaves, the Hebrew, Abraham and Sara abused their Egyptian slave, Hagar.

We watch as god invites Israel on to the mountain with them to be a nation of priests. But how the people see the storm gods cloud and are frightened. This is echoing a similar transition of deity worship transferring from el to the storm deity bal in Assyrian, cainanite cultures around the same time. And so we are told whyh bars them from the mountain, choosing only mosha to stand in gods presence. We see how Israel constructs an idol in the form of a golden bull, because this is the form of god they are comfortable with. And when mosha confronts Aron about this he responds by claiming that "they threw the gold in the fire and this was the shape it took". Essentially saying that this is the way it's always been and so who are they to question tradition. (The way god describes Israel here is as a Stiff necked or ridged people. Essentially the problem god has with them is that they are too conservative or traditional.)

Now apart from this movement from a pantheon of gods to a monotheistic conception of deity. This whole story is realized in the words of yet another Messianic figure named Jesus who claims that we are the ones who bare gods likeness and so we don't need idols to envisage gods image, because that's our role. Nor do we need these complicated traditions or cleansing rituals to make us holy. but lastly that we don't even need a sacred mountain or temple to worship in because we are gods dwelling. this whole story was about god enacting new forms of creation thru humans, essentially bringing about the kingdom of god or a heaven on earth. In fact in many ways the early Christian movement was more a kindred to atheism than it was any of the other religions of its day. In the way it elevated our treatment of others over any adherence to an orthodox doctrine or dogma and was more concerned with deconstructing our ethnic, economic, and social identities rather then solidifying our identity to an ideology.

In fact I would even argue that the dogmatic, nationalistic, egoic, god neiztchie claimed we had killed in (thus spoke Zarathustra) was actually an idol and image of god that needed to pass away.

And in many ways this image of god attested by those who claim to be atheists is perhaps a better example of god then that of many orthodox Christians, who's identity have been tide up in ideologies that hates, fears, or shames, anything that doesn't fit its mold. A mold which is actually more a likened to the idol then the ineffable concept of god.

god loves everyone, well except for homosexuals and people of other cultures/beliefs of corse! God loves every one so long as the look and think like us!

But here's the thing because being attracted to members of the same sex, is as much a choice as being attracted to those of an opposite sex, ultimately it boils down to the way we are made! The point is, we are required to love our enemies or at the very least abstain from hating them. And yet this small dogmatic god can't even have compassion and love for what they made! If god's form of morality is more abusive and oppressive than that of the most virtuous man, if humanities capacity for love and compassion has surpassed god's, then we have a problem. I don't care how mighty and powerful your god is. If humanity is the "bigger person" then your god is small. Jesus had to die In order to make humans acceptable to their creator. Because God lacked the capacity to love his own creation (which mind you, he made in his own likeness). This is Like a bad parent hating their child for being just like them. But This is fitting because the biblical god often utilizes some of the most abusive and toxic parenting tactics known to mankind. Tactics that have been proven to be ineffective and harmful in rearing children. The point isn't that god is evil but rather that the Bible is an ongoing discussion pursuing gods true nature. And so obviously we will get some very bad examples and hypothesis's pertaining to that god and nature in question. People will preach the god of the Bible says this, or that! but which one? Because there is not just one god in the scriptures. This idea of monotheism in the scriptures is about teaching us to see the world more holistically as opposed to dualistically.

The most frequently used title by Jesus was not the son of god but rather the son of man or basically the human one. The fulfillment of the tora was meant to make us more human. And so how is it that the law instead made man into a heartless monster that saw our humanity as something to cure. And even the law intended to lead us to god instead become an idol separating us from god. the question remains what is god? As far as the Bible is concerned perhaps the author of 1 John came the closest when they claimed that god is love.

And surprisingly this fits quite well with the name god gave themself when mosha asked "who shall I say sent me?" In exodus. This name, tho utterly unpronounceable in its original tradition is quite simply I AM or THEM WHO IS. And with it was this claim of omnipresence. This was not some dualistic name that separated god from creation, but a clam that god was present in creation. Because god's answer wasn't a proper name but rather presence itself. "I AM WHO IS" You see this was not some malevolent god outside of creation judging it but rather a god that was present in and permeating thru all of it. And what else would you call that but love. Recognizing that the world and all life is not in fact divided against itself but rather integrally connected.

And so when mosha saw the burning bush. It wasn't like the ground somehow became holy but rather that mosha was made aware of the grounds holiness. An interesting point here, because this book was originally written in Hebrew. and in Hebrew the word for ground is "ha adama" and the Hebrew word form mankind is "ha adam". And so it would have been clear to its original audience that we were the holy ground and to know god was to treat people, ourselves included as holy. Or more simply put, to love one another and be our brothers (and sisters) keepers. This is why It always bothered me growing up in the church when I saw the way Native American cutler was vilified because they were "heathens" who worshipped the earth and didn't subscribe to "the right" doctrine. But if you can't find god thru nature. If the rocks don't profess god, well then we've got a problem.

we get this account from Columbus upon first encountering the indigenous people and he describes them as the most innocent honest and compassionate people, eager to give any stranger the shirt off their back and the food from their table. and yet the very first thought Columbus has is that these people would be so easy to in-slave and indoctrinate.

And this is not even the tip of the iceberg when it comes to people being oppressed abused and even tortured and killed in the name of god.

And perhaps that's why it's important to recognize that there is not just one god in the Old Testament, but rather an ongoing argument all through, pertaining to how we distinguish between them. Which values are good and what examples are harmful.

Which brings us to job, because we get perhaps one of the best examples of god at work in creation in this story.

In this story we find a good man named job. In fact we are told that job was set a part by god in all the world. Not only was job a good man. earnest and studious when it came to upholding the laws and traditions so as not to offend god on his own behalf but also as an intercessor on behalf of his children. for job was a great man with great wealth. We are told that he had 7 sons and 3 daughters both numbers of which signify completeness and holiness.

The reader is also prevê however to events transpiring in the heavens among the members of the divine council where god presents job as an example of righteousness before the opposer. To which the opposer argues that job is only good because god had never let evil befall him, but if evil ever was to befall job he would most certainly curse god. And so the story goes that god permits the opposer by gods command to inflict evil or calamity on all job owns but as for job himself he may not harm.

All at once the reader watches as all of jobs servants, cattle and livestock, are reduced to nothing and further more we are told that a great wind came and desolates the four pillars of the the house in which his children were gathered, subsequently killing all who resided inside. And yet to all the calamity that had befallen job, he did not sin but praised god in his humble grief and obedience, blamelessly accepting his fate. At seeing this god praises job as one with great integrity before the council and yet satan persists claiming a man may sacrifice all he owns but if you allow infliction to befall his flesh then he will surely curse god. And so by gods hand and authority job was then inflicted by painful sores over his whole body but as for his life he was not to be stricken. By this point even jobs wife pleads with him to curse god and die. But still job remains faithful and dose not sin against god or man, exclaiming " should we except good from god and not evil"

It is here we are greeted with jobs friends who try and rationalize jobs predicament even proposing that job had wronged god, for god would not curse a man with out transgression.

These arguments attempting to rationalize the ways of the heavenly deities with the logic of the earthly man are quite extensive. But after much discourse finally god answers job.

In Job 38:2-38 (authors amalgamation)

God Essentially says to job "I AM the one who employed lady wisdom. I AM the one who built the rock, the firm foundation on which your house was built. I AM them who told the proud waves where to cease and put chaos in its place. And yet I AM in the chaos waters, and the storms who bring calamity go out on my command. For job I AM the LORD god of both order and chaos.

God continues In chapter 39:5-8 (authors amalgamation)

Do you see the stubborn outcast who will not conform to society but who wonders the wilderness. I AM with him!

Job 39:13-18 (authors amalgamation)
Do you see the fleeting fool who counts her labors as vain. Unmindful that her toiles were for nothing. Her fleeting spirit inspires me For she is MY dearly beloved.

Job 41:1-34 (authors amalgamation)
Do you see the ferocious chaos monster who no man can contend with. The one who's will can not be bent or broken. For I AM his confidant and we speak tenderly with candor. We have coffee on Tuesdays.

Here as well as all throughout the scriptures god claims ownership and presence in all of creation. (Isaiah 45:7) (KJV)"the god who created and works thru both good and evil".

(Mathew 5:45) NIV "god cases the sun to rise on both the evil and the good, the rains to fall on the righteous and the wicked alike".

But far to often I think we subscribe to this idea of a dualistic pantheon of gods when we conceptualize god as this malevolent dogmatic judge outside of creation rather then this compassionate presence working in and thru creation in all its dichotomy.

Another common image for god, especially in the Christian traditions is that of the trinity. Here we see the three distinct archetypes of the father, the son, and the spirit. These Archetypal images can be more accurately described as …

the father, the primordial source (el enlyn) the ancient one, the progenitor, or even perhaps the will of nature and fate.

The son or the human one as the image barer of god and partner in creation, as the one who both submits to the will of nature but also exercises their authority in the realms of traditions, doctrines, systems, and political laws.

And the spirit who functions as an intercessory between nature and culture, god and man.

Man is submissive to nature but not to culture of which man is the ruler.

AE "the sabbath was made for man not man for the sabbath".(mark 2:27) NIV

We are not victims to our policies but rather those who govern these traditions and ideologies. the human one seeks an alignment of will with that of nature, the process of evolution, or god. But is not concerned with conforming nature to culture.

it should be noted that originally the trinity was a concept constructed by the council of Nicaea to rationalize the divinity of Jesus in a monotheistic religion. An entity who is both independent from, and integral to god. Similar to the way we have a heart, a mind, and a body. And yet they are all part of the body. Your thoughts, your feelings, and your physicality are all extensions of you as an individual. And tho the theology surrounding this idea of a trinity is problematic. ultimately it dose serve a vital role in affirming that tho This world is diverse, it is not divided against itself.

how did a collection of texts, compiled over a period of 1,000 years and influenced by over half a dozen cultures, came to assert a god who is restricted to one cultures interpretation? How did a god who spoke for the outcast and marginalized, a god who incited justice on the abusive power structures of the ancient world, come to be a god who only loves those who look and think like "us"? How did a message about recognizing the very ground we stand on as holy, the good news that claimed we didn't need a sacrifice to validate our existence, then became the very abusive power structure which dehumanizes and demonizes humanity and the world we live on? These questions are of course rhetorical. the messianic figures we find in the Bible fought and argued with god on behalf of humanity) from Noah and Abraham to Moses and Jesus. These characters wrestle with god to be truthful and compassionate towards people. These elect people deliberately opposed the shame, fear, and hatred, of their cultures. In favor of compassion, love, and mercy, towards the fellow man. Ultimately they risked exile form their communities and rejection by their tribes traditions (the equivalent of hell in the ancient world). All in the pursuit of bringing about a heaven on on earth. And so how did a god who cared more about how we treated people in the here and now, become the god who is more concerned with our subscription to an egotistical ideology to deliver us from a cosmic hell when we die?

God? Not the egotistical personality!

god didn't write the Bible, man did.

why is it man's responsibility to prove god's existence? Is god not competent enough to speak for themself?

Why do we feel the need to defend god's honor?

The point is if god is all powerful and yet does not then (at no inconvenience to themself) materialize themselves to those who cry out to god, then how can god be good. What good mother withholds from her children love and affection, even at great cost to herself? And yet an all powerful father can not even be bothered with acknowledging the creatures he created completely against their will? You say gods word is in scripture and yet not a single word was directly written by god, well perhaps the 10 commandments, but not even Jesus kept all of them! (that is, he not only dishonored his father and mother but deliberately disowned them, (mark 3:33) not to mention working on the sabbath. (John 5:16)(John 9:14). ultimately the point is not the laws themselves but the context. Jesus recognized that the sabbath was made to serve man not the other way around and if it was In our power to do good we should. Now don't get me wrong, I want there to be a god. but this conception of an egoic persona is not only Not conducive for a healthy psyche but also lacking in congruity with the observable laws of nature. That dose not mean there is no god but rather that our conception of what god is, is seriously flawed both in moral application as well as testable observation.

If god truly was an ego, then god wouldn't have left their identity up to humanity to decide. But instead when Moses asked for identification, god did not give a name but answered with presence itself "I am"

Don't get me wrong, I'm not declaring war on god. No. In fact all this is done in pursuit of god. I'm declaring war on man's mislead presumptions of who or what god is.

You can still believe that the universe listens and the cosmos cares, without subscribing to an egocentric deity divided and disdainful against its smaller parts. prayer is the means in which the solitary creature confides in The universe itself as a friend. It's mere mortals befriending the ineffable and the chaos dragon also. Prayer is how individuals find solace in there solitude. To speak to life as if it cares about all its tiny parts. To believe that the ultimate nature of nature is like that observed in a mother who provides and nurtures her helpless babe.

I'm not requiring my reader to subscribe to this conception of god but simply to loosen their grip on the ridged idols that have been passed down to you by the traditions of your given culture and people of origin.

if an artist draws a picture and it ends up not looking like the artist imagined. Dose the artist burn the picture in an eternal flame or simply disregard it as trash. And furthermore who is to blame for the unfavorable outcome? Did the paper refuse to be drawn on, or the led settle in the wrong way? If a watchmaker makes a watch and from the very beginning the watch is broken then who is to blame? The watch or the watch maker? The pot or the potter? The work of art or the artist? If a hungry child stole a pice of bread, is it the child who was starved and hungry or the god who made the child with a hunger and put them in a world where bread is scarce, who is to blame. Dose it really make sense To condemn creation for the creators design?

Conclusion

we don't see just one god in the Bible but a plethora of gods. and the one god in question changes over time, or more accurately our conception of that god evolves from an ancient one (parental figure) to a storm deity or master. And yet through the whole story god keeps calling their people to be their partners and join them on the holy mountain. This is key because from Abraham to Jesus, the primary role of the messiah (those in partnership with god) was to challenge the current cultures assignment of heaven and hell, and to rectify the abuses of that religious power structure even at the risk of encoring the punishment for oneself. All so they might eradicate the earthly hell those institutions had erected thru fear, enmity, and shame, in the pursuit of bringing about a heaven on earth thru compassion, acceptance, and hope!

This was humanity challenging the small and abusive conceptions of god, for more larger and compassionate ones. These messianic figures made clams like god doesn't require human sacrifices anymore (Abraham) or god doesn't require animal sacrifices anymore (Jesus) and perhaps even latter claims about morality beyond mere obedience (neiztchie) or the universal interrelation of all life one earth (Darwin) or even the rejection of an ego centric universe by Galileo. These are all profoundly blasphemous statements about the nature of god and their creation. Because they each challenged their conception of god. Truthfully I don't care how mighty and powerful the "god of the Bible" is, if my finite capacity for compassion and acceptance transcends god's. I don't need some supernatural display or miraculous event, if god's nature can't be found in nature (who's nature by extension is that of god's). Going to heaven when we die is not worth erecting a hell on earth. I seek compassion and inclusion not enmity and duality. if god cares more about his ego being validated then he dose about how we treat people and the world, then that god is not only small but wicked and vile. If we must make ourselves small and even despise our very nature, in order to exalt the god who made us that way. Then we've got a big problem. Now I will point out there's a big difference between depriving us of our expressions of compassion and empathy, and stripping us of our ego and false personas. The first is abusive, the second is liberating.

Poem

how small must a god become to fit within the confines of our doctrine and dogma! Is man's capacity for love and compassion greater then that of gods? How then can we love our enemies if god condemns the ones who sought god, the ones god themself had made in their own image?

(you may think I'm being critical of god, but I'm not. the point is this god who is restricted to one cultures interpretation isn't the god who created all mankind, but rather the god who was created by man.)

"we're only human"! But in there lies the problem because we (by which I mean religion) have degraded our humanity to some despicably wretched things. but my question is, is how small must our dogmatic conception of god be for it to requirer us to make ourselves nothing and even less than nothing in order to exalt that dogmatic god to a position of glory. If your god requires you to make yourself small in order to elevate him to a position of praise, then your god is too small.

Quick note
before we continue I would like touch on a detail discussed briefly earlier in this chapter. My reader has probably read Isaiah 45:7 and would note that the word used there is not evil but rather calamity. and tho the word in its original Hebrew is in fact ra or evil, our modern concept of evil and that of it original audience may be quite different. To the modern reader, evil is not something observed or exhibited in nature but rather a perversion of, or divorce from nature by mankind and thus not an term used to describe natural phenomena. However a point we often forget especially in philosophical or theological circles is that how ever estranged it may appear at times, human nature is in fact an extension of nature. And tho genesis 1 distinguishes man as set apart from the animals. Genesis 2 asserts that we are in fact also among the animals. and so

even the biblical account of creation are in unanimity, and most certainly not in contention with the scrupulously extensive evidence that man did in fact as Darwin proposed evolve along with all life on earth from more primitive forms of sentient beings. And it is my belief that nether the discoveries of science nor the profound wisdom of scripture in any way presume to invalidate ether one or the other in their claims.

There dose appear in the fields of art and science anyways this inevitable gravitas or motivating force pulling, inspiring, and calling life towards more complete and compassionate states of consciousness and this dose seam to advocate for some form of intrinsic entity or spirit prompting and propelling life. However if this force is truly higher and wiser than mortal man then it would not be driven by dualistic ego or delusional persona but rather seated in transparent pursuit for truth and connection. And for this reason it is important to remember that nether the prerogatives of our genes (instilled in us by - - -) or the paradigms of our psyche (inspired by - - -), nether impulse nor idea are evil but both intent on progress and preservation consecutively. Nether the heretic nor the governing authority are the villain but rather crucial roles both of which are paramount in importance. And this is highlighted in the intrinsic necessity for rest and sustenance to perpetuate life. For it is this mechanism of depleting energy and recourses that produces motive for progress and posterity to form more profound states of life.

To believe the god who calls us to love those who don't look or think like us (our enemies) then lacks the capacity or compassion to themself love those they (god) has made to look like them (god) is ridiculous.

Ultimately god has left their character to be decided by humanity. That is if we believe the universe is unified and one, that the cosmos is compassionate and nature nurturing. Then so must god be. But if we believe in a malevolent master who incurs an insurmountable cosmic debt on creation simply for existing as it was made to. Then that is the image we shall bare.

CHAPTER 4

THE SOUL

A quick quack at quantum physics

On an subatomic level virtually all mater is energy!

Matter functions both as fixed particles and as fluctuating wave or frequency at any given time. And so when observed,

The nature of any given atom has the potential for both.

These terms of fixed particles and fluctuating waves are mutually exclusive and dualistic, so this unity of the two should be paradoxical.

The reason mater appears solid and fixed is because the atoms that make up the molecules for any given object, shares electrons with other atoms within the same molecule.

It is those same electrons in the atoms of one molecule that then resist electrons form other molecules that then solidify each's exclusivity and solidarity. The electrons in each are virtually indistinguishable from one another, that is apart from their association or corresponding relationship with members of the same molecules.

Matter is solid because the nature of electrons is repulsive towards one another and yet when those same particles form a bond, they form a strong one.

This bond however is not the rule but the exception. In most cases opposites attract. Electrons attach themselves to neutrons. And this relationship between corresponding electrons and neutrons and that of those adjacent to them then forms an quantum field. This quantum field or relationship between particles is what dictates the atoms identity, that is rather it makes up oxygen or hydrogen etc….

And so all matter is not only energy, but the nature of that energy AE quantum field (relationship between neutrons and electrons) is what defines the state of that matter.

The body and the spirit

It has been said that we are not our body but rather our soul. And that we simply have a body for a brief time. And tho I understand the sentiment, I disagree with its conclusion.

This objection is solely predicated on the definition of a soul. So what then is a soul?

Let us start with the spirit, because it is often equated or synonyms with the modern readers conception of a soul. That is this idea that the soul is strictly spiritual or ethereal and dose not pertain to the caporal state.

The Hebrew word for spirit is "ruoch". This word in Hebrew also means breath which is similar to the Greek word for spirit "numa" which also means breath.

So the Greek word for spirit is numa.

The Hebrew word for spirit is ruoch.

And they both also mean breath.

"The wind blows wherever it pleases. You hear its sound, but you cannot tell where it comes from or where it is going. So it is with everyone born of the Spirit."
(John 3:8) NIV

To its original audience the invisible force of the spirit was equated with that of wind. and so often throughout scripture god's presence or voice came in the form of a great wind or gentle breeze sometimes even depicted as silence or still and yet at others as a ferocious storm.

And even today the agreed pronunciation of the name of god or "whyh" is in fact the sound of breathing.

Breath is not only vital for our survival but the way we breath can greatly effect our recovery when exercising, by efficiently distributing oxygen to our blood.

In fact slow deep breaths are often used in practices like yoga to regulate our nerves.

Prior to the late 18th century the belief that evil spirits were the primary cause of ailments was the most prevalent and prevailing conclusion. It was not until 1876 when one of the founding fathers of microbiology (Robert koch) pioneered the discovery and study of germs, connecting them with diseases and pathogens as the primary cause of infections. along with notable figures such as Louis Pasteur (co founder of microbiology) and Antoni van Leeuwenhoek who is noted with discovering bacteria as early as the late 16th century. This belief that diseases and infections were purely spiritual began to fade with these discoveries. However before these discoveries, things like the quality of the air we breath and the presence of pathological pollutants was associated with forces of an ethereal nature rather then to those present in the physical world.

However this word spirit encompasses far more then strictly the pathological and subatomic world but more commonly as it is used today to describe the psyche or our psychological state. We use phrases like she is high spirited or his spirits are low to describe our emotional states. Which is understandable because our emotions are fine tuned for detecting abnormalities in our bodies such as diseases and ailments. Not to mention that positive emotions have been proven to be effective at combating diseases and strengthen our immune systems.

And so it's no surprise that the most common concept of the soul is simply inflating it with our spirit as a strictly ethereal essence. however when we do this we reduce it to less then half its intended role.

This misunderstanding is most likely due to its roots in the Greek word psyche. And in accordance with its Greek context it is most certainly understandable why our concept of a soul is entirely metaphysical or even perhaps philosophical in nature. Now there is no doubt this concept heavily influenced

The later usages of the more ancient Hebrew word for soul " nefesh" which did come to mean something more a likened to a strictly spiritual essence. Tho it's original usage was much different.

Prior to the Babylonian captivity in 560 BCE the most common usage of this Hebrew word "nefesh" or soul, was living being. This word is also used in depicting a neck or throat as that place where both body and spirit are joined or united. not merely as one's spirit but as an entity dependent upon the equilibriums relationship between our bodies and our spirits. This is gracefully illustrated with this imagery of man being breathed of dust or a ubiquitous union of the physical and the ethereal. A sort of heaven on earth. Where the ones made in gods image were as god was, for god was the scours of the heavens and the earth, and as an completion of creation god joined them once again in the form of humanity.

Heaven and earth in one body. And so once again we see this repeating theme of the raw undefined whole being split into parts that can be refined. all with the purpose of being rejoined into a more complete whole. We see this with god taking a part of man's side to make woman, but with the ultimate intention of the two once again becoming one flesh.

And so there is the whole, which includes all the parts. And so regardless of how diverse the dichotomy of those parts may be, they are not mutually exclusive.

Now we often hear the argument that our bodies are temporary but our souls are not. But exactly what part of our soul. I mean yeah our bodies decay and deteriorate, but so do our minds. And so what part of our soul is exempt from afflictions like Alzheimer's and dementia, not to mention conditions like ADHD and autism. Exactly what part is our soul, if it isn't concerned with our physical or psychological characteristics? The point is none of it is permanent. And all of it is connected. The soul is not one or the other but both. And it's the fact that it's not stationary or static but rather fluid and flexible that then allows for its entirety to be expressed.

that's the cool thing. Because regardless of rather we want to or not, we are continually ether rediscovering or redefining what it means to be us, in this moment and in this phase of our life.

this is what's depicted in the Chinese yin and Yang. As they become aware of themselves they see each other clearer and ultimately find they are both distinct and yet at the same time one.

"We are what we repeatedly do. Excellence, then, is not an act, but a habit."
- Aristotle

I am inclined to the view that things were generally done first and that it was only a long time afterward that some-body asked why they were done.
C.G.Jung

Our physical habits and practices play a huge role in our mental health and it is often in doing small ordinary tasks that we stimulate our synapses and cultivate neuro pathways for healthier ways of thinking. Or in other words, more often then not an healthy body leads to a healthy mind and this is partially due to our brains getting the necessary oxygen and blood flow to work efficiently. And this is why it is important for us to recognize that we don't simply have a body but that our body is an integral part of us as a whole.

Relationships

all love is conditional, but sometimes that condition is you!

more often than not the expectations (or conditions) for attaining love are predisposed by our parentage. To some degree these conditions are propagated genetically. however some of them are ethical and thus malleable. Which is good because it is this criteria which exemplifies not only our standards for potential mates but more profoundly ourselves. Our definition of love is predicated on our own expectations for ourselves. This makes sense because love is often depicted as seeing oneself in someone else. And so when someone cheats or betrays our trust our schema is challenged and even contradicted. That's why we say things like "I thought I knew you" and " I don't even recognize you anymore". The problem is (and this is most certainly not a problem) people are not stationary objects! And thus we are constantly rediscovering, and if we are wise redefining what it means to be human. And tho love is most certainly seeing ourself in another, it's also so much more. It works both ways. Not only do we see ourselves in others but we see others in ourselves likewise. And that's the part that scares us. It is also what inspires us. because it articulates that love is not limited to our conditions. And so all love is conditional, but sometimes those conditions are your ability to tolerate and even appreciate people as they are.

in virtually all ancient cultures, women were considered as property and were bought by the groom from the patriarch or father. Thankfully Over time women have attained their autonomy in their societies and cultures. and yet it is still the custom in the west, for the man to pay monetarily for their interactions when courting a potential mate. rather that be paying for dinner, entertainment, or venue, the relationship still in many ways takes the shape of transactional.

Now other animals do this to, however the primary reason animals expend such recourses is to obtain a reproductive mate. The point is this objectification goes both ways. A partnership is about two equal party's sharing their time and energy with one another because they enjoy each-other's company and not as a transaction to obtain offspring or monetary security. the problem with this mentality or state of mind is that it predicates this idea or identity of ourselves and others as products to be bought or sold. And so instead of simply being what we are and appreciating others for what they are. We end up branding ourselves and others in order to validate our worth as a person. We ask what can I get from this person with my resume! Am I pretty enough or rich enough, successful and talented enough, smart enough or funny enough to earn that persons affection. But that's not love! Love is not about getting something! It's about appreciating what is, as it is!

Now it's ok to have healthy standards. You don't need someone who demeans or abuses you! And if you don't admire someone for their expression of life. Most definitely don't settle. Because it's not fair to you. and it's not fair to the other person. The point is, you are not a product or a utility. And a persons worth is not tied up in what they can do. but rather how they chose to be.

Ultimately I believe this practice of courting and dating is Ill equip for producing a genuine relationship and partnership of any real magnitude. And instead It is thru the unconditional nature of friendship which life produces an environment where people can grow together without the baggage of presenting a polished persona or product in the form of funny/charismatic façade, physical charm/beauty, or potential for financial stability. This transactional environment of courting dose not cultivate conditions conducive for a congruent connection. However because of the lack of mystery among friends the crucial component of thrill posited thru flirtation and it's subsequent energetic attraction is often defused. It is this lack of fear or uncertainty that impoverishes a friendship of plentiful potential for chemistry or spark. Regardless in regards to partnerships intent on depth and longevity the transparent environment of a friendship is far better at cultivating meaningful connections. It's no secret that dating engages the same mechanisms in our genetic coding that propagate a predator prey dynamic as opposed to a partnership. And yes it is that dynamic which sparks our thrill and focus but it often devolves that same goal in the ingenious signals and disregard for the value posited in the individual. That is what this form of relationship often results in is one party pulling away out of fear (like prey) and inadvertently triggering a reaction from the other party to ether pursue more vigorously or pull away also (predatory) This phenomenon is often described as ghosting by at least one of the two parties and more often then not results in the termination of the relationship. In fact the only response which really proves to be effective in this particular circumstance is to do nothing. which can be a display of solidarity and emotional integrity but also can assert a lack of genuine interest. This is very problematic because in the first case it elevates one party to a position of some sacred object and the other is downgraded to no more then a champion in pursuit for this affection. And in the second there is no real evidence of a connection at all but simply an acquaintance of convince.

You can't love someone else until you learn to love yourself. In much the same way, the way we treat our body is the way we treat our minds by extension. And thus the way we love our selves is the way we will love others. Or perhaps more accurately we show others the love we ourselves crave or believe we ourselves need. And so if we believe we are broken things that need to be fixed we will then try and fix someone else as a display of our love. However if we view love as merely a social construct for alleviating loneliness and elevating one's position in society, one may attempt to prove their necessity or utility in another's life as a means of proving their love. But this not only objectifies ourselves but by extension our partner as well. Because the truth is we are all trying to love one another as best as we are able. the problem is we are often taught that love is this thing we get or obtain. rather than something we already have. much like ourselves and life in general, love too is not static or some stationary object. but rather and ebb and flow, ever changing shape. And so the point is not to obtain it, but to resonate with it, as it is! in each new moment. It's like a song or a wave in the ocean. because it's about being connected and in sync.

This is the inconvenient truth about our souls. Because if our hearts are not in it then our bodies won't be ether. And if our bodies are distracted because their needs aren't being met, then our thoughts won't be able to focus ether. If we are busy trying to make what is, something else! we will cut ourselves off from experiencing what it actually is.

The flesh

Now I know there is a lot of animosity often directed towards our flesh or bodies and we can have a tendency to reprimand our bodies when our nerves or emotions are not functioning they way we won't them to, or more often then not the way we don't want them to. Because we've been conditioned to believe that the majority of our emotions and nervous responses systems are negative or inconvenient. When in all actuality are bodies are finely turned to picking up on our environment and telling us what we need, or alerting us to any potential dangers that might be present. And so the point is most certainly not to permit our emotions to rule us, but it also is not to invalidate our emotions ether.

an useful tool when dealing with stress and anxiety is exercising or simply exhausting our built up energy which then discharges the chemicals our bodies produce to protect us and permit us to escape natural threats in The wild. This can be done in any number of ways, from literally exercising, to practice like yoga, dancing, wiggling, shaking, walking or hiking, or even playing sport. The point is our bodies are not the enemy and we should be learning how to listen to what they are telling us and then being able to communicate back to them. Because we are not one or the other but both, our bodies and or minds. and we should be striving for equilibrium rather than omniscience of one to the other

Our bodies are how we interact with the physical world. And so without our physical extension our minds and hearts would have no way of bringing their ideas to realty and likewise without our minds, our bodies would be devoid of inspiring imaginative forces to propel it.

This body is yours! And so is this mind! you are not one or the other but rather the relationship between the two.

Our identities are not in one or the other. The meaning is in the relationship between the dichotomous parts.

in its intended context this claim of being breathed of dust had more to do with being both physical and philosophical creatures. Rather then depicting this pull to the darkness or light, predicated on our bodies or flesh being bad and our minds or psyches being good. Ultimately this deifying of our existential quality wile demeaning our physicality, is not healthy.

It's important to state here to that our bodies are not bad, our emotions, nervous systems, and bodily impulses are vital for our survival. And so what ever it means to be you, here and now, in what ever shape that takes, be that! Because this world is big enough for all of you, as you are, here and now.

Yet still as is far too often the case we feel a need to degrade our body or physicality in order to exalt our mind or spiritual essence to a position of sanctity. we fear that if that hierarchy isn't sustained. We will lose our meaning entirely.

But this is not the case. The meaning is not allocated to one or the other but rather the relationship between the two.

Here's an example.

When you kiss someone, the action itself is arbitrary. and yet if you like the person you kissed, the kiss then means something.

Or for instance take food and sleep, they are ultimately mundane and trivial, until your hungry or tired. It's the relationship that gives these things meaning and not the things in themselves. It's The desire and lack, that produces value and further more meaning. Much like the atom. We too are defined by the relationship between our parts.

it can't be over looked that to some degree people are a result of their experiences, and still even more subject to their chemistry and genes. And yet there is still something that transcends ether variables attempts to dictate our fate. There is some deep calling often disguised as destiny which is relentless in defining our identities. And rather that can be broken down to our predatory instincts proclivity to fixate and focus on a target or quandary, or some divine intervention. There is an unquantifiable quality to humanity.

Ultimately In much the same way the philosophical is an extension of the physiological, the mind too is a manifestation of matter and not a separate entity In duality. For mater produces the material for the mind to manifest. There is no software without hardware, no program without circuitry, no psyche without neuro pathway.

Quick side note. This idea that discipline is about completely renouncing and denying our desires is completely missing the point. Because the role of discipline is to stimulate those muscles and feed that hunger. That's exactly how flirtation works, as well as exercising specific muscle groups. You see because instead of quenching the urges, discipline tempers that hunger. We stretch our capacities and tolerance for things like pain, subsequently intensifying that desire and multiplying the potential pleasure and relief posited when those urges are finally meet. These are the ultimate goals of fasting and celibacy. It's about attributing value, not degrading it. And this is the problem we shall be discussing in the next chapter. Because just like denouncing certain urges leads us to vilifying them, the very same disciplines can often lead to those same urges being exalted to a position of idolatry. This is a common phenomenon observed in those who retain their virginity and subsequently end up both cultivating enmity and adulation for neutral urges that are nether inherently good or evil but nun the less vital and crucial.

All this to say, rather it is pertaining to obtaining sustenance, rest, or sexual attraction, these urges are not bad, in moderation.

CHAPTER 5

GOOD AND EVIL

"I would rather be whole then good"
–Carl Jung

One of the pivotal questions in this discussion may very well be, is man inherently good or evil?

This highlights both mans propensity for corruption and decay, wile often forgetting that it is that very same trait that grants us our capacity for adaptation and growth. But after weighing the horrendous acts of human history with those deemed virtuous dose the good out weigh the bad?

According to religious sources the despicable trait inherent to humanity is sin.

But what is sin?

Because throughout the century's sin has been the justification for religious authorities condemning, shaming, excommunicating, and even executing, the accused. And yet according to Luther we are told that sin is something we inherit from our fathers from the very beginning. So how can we be guilty of something punishable by death, for simply being born. we didn't even choose to exist, and yet we are guilty! Guilty of what exactly? Existing! Was that our big crime?!

You see how ridiculous this is right?

If god made us. And we are inherently riddled with sin! Then god is to blame!

But you see originally sin wasn't inherent to humanity. The Hebrew word which we most often translate as sin is "hata" which simply means to go astray.

originally sin was a word made up to describe humans rebelling agents the natural order. However this rebellion often included our natural impulses. This is the big evil monster we call sin. It's coding in our genes. Things like predatory instincts, faulty or intentionally misleading forms of signaling, reproductive imperatives, and even includes attraction to certain traits. Now don't get me wrong. We shouldn't go around killing people or stealing from people. But to say it's a sin for finding people of the same sex attractive or having sexual impulses towards people of the opposing sex, when those things are in our genes and not choices we made. The idea that we are judged on a cosmic level for things we had no control over (like our biology for instance) is absolutely absurd!

Murder is wrong, theft is wrong, rape is wrong, deceitful signals understandably warrants distrust, prejudice and human slavery is absolutely despicable (and yet the Bible not only condones slavery but also commands genocide) the point is the Bible is a bad standard for universal morality at this day in age.

Now as we've already discussed mankind has an irreparable desire to impose a form of culture to govern our nature. and understandably so, due to the chaotic nature of our intuition and impulses. But also as we've observed it is often that very imposed order that results in the perpetrating of the most horrendous atrocities such as that observed in the supreme order of nazi Germany or soviet Russia.

One of the most universal qualities in practically all sentient life forms is our curiosity. This transcendent desire to puzzle and play. This drive to engage with our environment is observed in virtually all conscious creatures. And yet it is that very same impulse that often has the proclivity to leading us into trouble. Tho I shall also point out that it is that same impulse which is responsible for solving the majority of our problems.

there is some profound Talmudic literature that offers a commentary on the claims of genesis 1, that can elaborate on this very idea. It accounts the discussion of Rabbinic sages who claim

"that tho the prognosis of good attributed by god to creation in the first 30 chapters of genesis is referring to the pleasurable things. the very good then in verse 31 is in reference to those things we would call evil or bad. and that if it were not for those toxic or destructive impulses like greed, pride, lust, and even fear and anger, we would not then be inspired to create, be united with another, procreate, or conduct business of any kind. Not to mention enact justice for those wronged."

"do not call anything unholy, when I GOD have created it and made it holy"
Acts 10:15 (NIV)

In fact the first time god ever refers to anything as less then good, it is the state of man's solidarity.

"It is not good for man to be alone"(genesis 2:18) NIV

It is here that god causes man to fall into a daze and takes from man's side to create woman. Now this word Adam in this context was not strictly the name of one man but rather analogous with mankind as a whole. And so as we've already discussed god takes a part of the whole to set apart as an "Azer" often translated as helper in this verse. however elsewhere this word "azer"takes the form of savior or more accurately as the one who brings salvation. In fact in Talmudic literature man is not a whole person until married with a partner or wife. This phenomenon is expressed in the ancient Greek culture in the form of a soul mate. And it bolsters the idea that we need something or more accurately the objectification of someone to complete us and make us whole.

Tho much can be said about the procreative purposes of these teachings (that the two become one in order to create new life)

The more important point here is not that we need someone to complete us (which we do not) but that complete isolation from any form of dichotomy is most certainly not good. That traits in omnipotence (rather masculine or feminine) are not inherently good or evil but dependent more upon their relation to one another.

I would like to make a point here and that is. Marriage is for here and now! (Jesus eludes to this in In (Matthew 22:24-32). Or for instance take the marital vows "tell death do us part". Marriage has no cosmic value According to scripture).

The misconception of duality between good and bad.

It's funny when you think about it, the things we perceive as bad often have far more to do with their relationship with us then they do with the autonomy of the object itself.

For instance cannibalism is generally thought to be abominable. however it is perceived as redemptive for the king snake who devours its venomous brothern, it's also perplexing perhaps how the common defensive traits of herbivores creatures such as hooves and horns are associated with demonic imagery. And yet We do not vilify the predatory instincts of the lion, and even admire its domesticated progeny as a cute kitten.

We understand that animals prowl, procreate, and kill its prey, or its predators in defense, all instinctually to survive. And yet because we're conscious, we feel shame for those instincts in our selves. (Now carnivorous creatures obtain far greater nutrients from small portions of meat than herbivores obtain from massive quantities of vegetation. and so it's no wonder why humans evolved predatory traits. It was those very same genes which allowed us to allocate the necessary recourses for our big thinking mind.)

This dose not mean it's ok or in any way conducive to hunt and kill. However those instincts are still present in our genes and we are still predatory creatures. However those violent behaviors are not conducent to procreating life at this stage In our evolution. Sex on the other hand is still necessary.

The early church, or Luther anyway believed it was that particular habit that instilled original sin in humanity. And thus by simply existing we were inherently evil as a species.

And so not only are we incomplete on our own but also irrevocably vile because of our impulses to procreate. It's interesting how we have shame for any display of nature, and idolization of artificial laws such as doctrine and dogma which have become signs of piety. Because heaven forbid we should learn to admire and appreciate yet alone even love the world we live in and cease our chastation of it. I have found we tend to prefer the gods we have erected in the words of our books. rather then that expressed in life's nature, for which by extension would be that of its creator.

As if anything of nature, wild, and rurally, is evil or at the very least taboo. and yet any abominable propriety constructed by man is worshiped as sacred and holy. However it is no secret that we naturally feel shame or embarrassment when we discuss the subject of sex. But this is not because sex is sinful but rather because it is intimate and leaves us vulnerable and exposed.

(I'm not saying sexual immorality is the only sin. but most religious traditions tend to be far more concerned with whom and in what manner it's members associate, then they are with systemic prejudice and injustice. they have no problem with things like gluttony and pride and are often found condoning forms of abusive behavior to those they deem "lesser". Not to mention all the killing and ceasing of property perpetrated by religious sects against those who spoke out against their injustices and made claims they deemed heretical.)

Biblical Plot holes

sin is often equated to a disease. And yet those labeled as sinners aren't met with the compassion of a sick person but with shame and enmity. However perhaps a better analogy would be that of the over consumption of unhealthy foods. Because like the "corrupt object" in question, things like Sugars and fats are not bad and are even necessary in moderation. the reason this analogy is perhaps more fitting, is because as C.S.Lewis described it, "sin is the good and pleasurable things god made, just out of their intended context". I find this analogy particularly fitting especially because it acknowledges sin as an affront or abuse of nature, rather than something inherent to mankind.

Interestingly enough Jesus didn't shame the prostitutes and tax collectors. however he had no tolerance for the money changers in the temple. For those who had turned the religion into a business, and monopolized on peoples misgivings, Jesus had the utmost distain for. Because if someone truly understands sin in its intended context they know it's more a kindred to trying to fit a square block in an round hole, than it is a debt incurred. Its a tool or recourse which was precisely designed for a crucial purpose. However because of certain circumstances, that intend outlet is obstructed, and so it exhibits its role elsewhere. But like the square block, that's not where it goes and so it won't fit properly. not only dose this usurped authority undermine the objects role, but also over extends its application beyond its intended capacity. And so not only dose this precise tool end up not being used as it is intended. but also by extension it is deliberately used inaccurately. Or returning to our original metaphor. Sin is both the greedy person who ate too much, and the hungry person who's food was stolen and thus must result to eating trash in an attempt to quench their hunger. In the end both parties are sick. And That right there is the common misconception about sin and vice. Because you can't just quit a harmful coping mechanism without first resolving the issue it alleviates. You have to find a healthier alternative before you can get rid of the harmful habit. Because ultimately The hunger must be quenched.

This is the most accurate illustration I have found in regards to what sin is and why it exists. And tho this answers the question in regards to the act of sining. It doesn't resolve the supposed inherent or inherited aspect of sin. But if sin is inherent, or the fundamental predisposition of humanity. Then sin is not just a natural response to an unnatural problem.

However if Sin is not an affront to nature, the fault then lies on the watch maker and not the watch itself. And so then if god is so disappointed with his creation, then why did he create it? And who other than the artist is to blame for shortcomings of the masterpiece? The point is god isn't displeased with their handiwork. All Throughout genesis and job god asserts their love and even admiration for their creation.

This idea of inherent inadequacy is a tool often used by religious institutions to necessitate their intercession. "If you don't subscribe to our prescribed interpretation you'll end up in hell"! "Jesus died for our sins! So accept him as your savior or burn for ever in hell"! But there lies one of the big issues. Because exactly what did Jesus deliver us from? we are all still subject to the physical death of the body 2,000+ years later, we are still required to die to our archetypal identities (John 3:3), and Jesus didn't come to deliver us from the ternary of the law. (matthew 5:17) NIV

("I came not to abolish the law but to fulfill it. Now This is a common euphemism used by rabbis to indicate that their interpretation of the scripture or "yoke" was the accurate one.) but if the institution of the church is the best example of what it looks like to be spiritually alive, then the saved is no better then unmarked graves. The only conclusion for Jesus's redemptive claim this day in age then, hinges solely on deliverance from some eternal cosmic torment (which didn't even conceptually exist until the 4th century when Augustine of hippo popularized the idea of hell as we know it today) before then this idea of hell was radically different. From the Greek Tartarus and hades to the Jewish gehinnom and Sheol, the underworld was simply a grave where the dead were forgotten by the living. According to antiquity, the afterlife was much more a kindred to the modern atheists conception of oblivion then it was the modern Christian concept of eternal torment. But this idea of death isn't scary enough to detour certain evils. Now I will admit In some early conceptions, hell is designated as a ethereal prison where those who had assaulted or molested family members, those who were dishonest with patrons, and those who propagated poverty by hoarding wealth and abusing the destitute, were then punished for their injustices. Although more often then not this had more to do with the respect of the community then it did a physical punishment. Ultimately hell was the denouement of an individual from the natural order as a response to that individual's transgression or manipulation of that natural order. Ultimately the point is there is no crime perpetrated in an finite lifetime that would then warrant an eternal punishment. Especially when that crime is a result of a predisposed prerogative in our genetic makeup, AE per the designs of our creator. It's like perpetually burning a grandfather clock for chiming every time the big hand hits 12! Originally hell was desegregated for the abusive institutions and those who had manipulated them to profit off of the marginalized. Unfortunately this idea of hell has been repurposed by those abusive religious institutions for that very purpose.

So Where did we get the idea that we needed a sacrifice to atone for our sins? (in the exodus myth this sacrifice of the first born is required by god for the transgression of political/religions abuse of humanity. It is this abuse which according to exodus necessitates the intercession of a Passover lamb who's blood is painted over the door of the israelights house or refuge (symbolic of the ark) this is significant because this final act of god (12th plague) is synonymous with the decreation of culture or order expressed in genesis by the chaotic flood. Interestingly enough this is also the role of The test! It is us passing thru the chaos waters, dying to our restrictive world views, and greeting the paradigm shift with compassion and grace.)

The transgression here was the institution which had abused nature and humanity by extension.

This sacrifice redeemed humanity by subjecting that utilitarian order to the rule of nature. Which often looked like chaos to those who had profited from the abusive systems. This sacrifice atoned for the transgression caused be the deification of an idea or usurped authority by the heavenly being. And not that of our animalistic impulses.

And so in this context sin According to nature would be expressed as follows ….

(Biologically, the primary prerogative of sex is to procreate healthy offspring)

(death is natural, murder is not)

The idolization of objects and ideas and the subsequent vilification of innate biological prerogatives is unhealthy and unnatural. Thus curiosity and admiration are pivotal for growth. idolization and enmity s how we stagnate and isolate, both of which signify and inorganic form of death).

You see originally sin was not some cosmic debt we incurred but simply a transgression against the natural order. Thus Religion started out as man's early attempts to understand the natural order. But The evangelical retort to this idea is that we need an universal code for morality. That if it weren't for the omnipotent LAW! people would go around killing, stealing, and raping one another. But this argument is as old as time itself and it is always predicated on the premises that "our standard for morality is ideal" but you see if sin is an affront to nature. then we need a morality that is in accordance with nature and not one complicit in the injustices of culture.

But who's nature should be the standard for morality? Obviously the standard should be one that benefits those who look and think like me as opposed to my enemy! One that works for us but not them? I mean Am I really responsible for a strangers well being, am I my brothers keeper? The question of who's standardized morality is ideal, is like asking who's language should we standardize.

However both the Evangelical idolization of virginity/ vilification of sex, and the secular desensitization of inter corse, express a very skewed idea of sin and morality. And most certainly one that is not in accordance with nature.
"well what about Jesus! He's a good moral standard right!" He's Completely blames and pure! So obviously Jesus would make a good judge! but what about the time the sinless Jesus

used a racial slur to address a cainanite woman in matthew 15. Now the Greek word here kynariois(small dog) in its original context wouldn't have carried the same derogatory sentiment as it would by today's standards. However he still dehumanizes the cainanite woman based on nothing other than her ethnicity. And at first he doesn't even acknowledge her cry's for help. According to Jesus in matthew 15 Jesus was not sent to save all mankind but only the Jews. And this is reinforced by Jesus on many occasions. However then again according to Jewish law, this atonement for sins was not required for gentiles. And so according to this passage, is god cosmically racist, or only the god of Israel? obviously seeing as how the story concludes with all people groups being saved, the answers is probably no! God is not racist. This has more to do with the promise god made to Israel that Gods blessing to humanity would come thru the line of David. And so It's worth noting that Jesus's claim here is probably more of a metaphorical illustration meant to convey a message rather then a historical claim. (Because An important detail when reading the New Testament is tho Jesus was most certainly an historical figure, The historical authenticity of the New Testament is very problematic. Aside from its numerous contradictions and discrepancies between accounts and the understanding that it's original writers believed that Jesus's return was immanent. The writings of Paul are the oldest in antiquity in regards to this collection of books, and the gospels were not written until 20 years later.) Paul's writings are dated to around 50 AD. Roughly 15 years after Jesus's death. mark, as the first gospel written, is dated to around 70 AD. And is directly sourced in the later accounts of matthew and Luke, John didn't come to fruition till much later around 90 AD. With this in mind we see things like the virgin birth, and the divinity of Jesus, slowly evolve. these traditions didn't just manifest in some divine encounter but evolved slowly and intentionally over a period of decades as an attempt to rationalize this agnostic movement of the way. But that's how life works, it doesn't just materialize fully formed but rather evolves and adapts slowly over time. The thing is we don't really need the historical authenticity to be valid in order to appreciate the moral implications of these claims. But even the figurative virtue of Christ is problematic and as for the historical ethic, we quite honestly can not tell. For even in the metaphorical writings of matthew there were places where the all powerful sinless Jesus couldn't preform miracles because the people there knew his past. (matthew 13:57-58)

Not to mention that the Unforgivable sin (mark 3:29) blasphemy agents the holy spirit, is the same crime Jesus was condemned to death on a cross for by the religious leaders of his day.

Note
This nationalistic idea of a messianic figure which Jesus claimed to fill, was ultimately an attempt to legitimize the promise god had given David " that one of David's house or offspring would always preside over Jerusalem (rule forever)"

The point is not even Jesus stands up to the scruples of today's religious oppression. Which honestly makes since when you realize just how cynical Jesus was towards the religious percepts of his day. Even stating he had come to set the world on fire and it was his desire to watch it burn. (Luke 12:49) or immediately following this when Jesus declares he did not come to bring peace on earth but division among members of the same family.

Now I'm not trying to undermine the moral impact of this influential religious leader.

It's no doubt Christ's influence has done far more for humanity then the author of this book ever will. But even his morality is not without fault. Jesus is not the devil, but he's not infallible ether. This need to deify Jesus is completely counter intuitive. The whole point was that Jesus was human, we don't need this sinless celestial being conceived of the Holy Spirit. The point was that our humanity was beautiful and holy and not some derogatory curse. This god incarnate was not meant to undermine humanity as image bearers. But to articulate that "whatever you do to the least of these, you have done also unto god".

I believe Jesus desired to be emulated, not worshiped. that is anymore then any person is worshiped, say a husband worships their beloved or a new borne baby worships it's mother and vise versa. Jesus's chief goal here did not appear to be obtaining glory but in conveying The defining traits of humanity to be desired or aspired to as those of compassion and fluidity rather then our egotistical rigidity and resignation to an idea. The point being growth in appreciating those qualities in humanity which had been posited by nature as opposed to those ideas exulted by culture.

You can see how problematic this conception of sin as a cosmic qualifier is then, right? Not only dose it make no sense. but it is also extremely harmful. It turns these ides of an absolute good and absolute evil into standards which make mankind subservient to the abusive institutions or systems. Rather than resolving the injustice and inequality, it propagates the problem. Ultimately This idea of inherent sin and its subsequent need for an intercessor to redeem us, cultivates an inability to love and accept ourselves and others as beautiful and whole people. This propagates this inequality of sin and subsequently prohibits us from seeing ourselves and other as any more then broken problems that need to be fixed and saved, or as sacred object required to obtain justification for existence. This idea of an inherent debt incurred by existing just propagates a mindset of scarcity.

According to this inherent standard of sin,

We are guilty of everything instinctual in us but not allowed to take credit for our strides at negating and even overcoming those instincts!

How ever there's a point here. because in ether case there are pros and likewise cons respectively to ether culture or nature as supreme judge.

it is the admirable qualities that make the inadequate ones abominable. For instance It was the discipline that made the nazis prejudice effective. But perhaps it was that prejudice that lead to its downfall. perhaps even, morality is not simply moderation. but rather that virtuosity is the most comprehensively efficient approach.

In ether case however, it was not human desire that propagated evil but rather indoctrination to prejudice, ideology, and fear of rejection. And thus obedience is not and can not equate to morality! It comes down to health. Why would a creator who made all of creation, then limit our humanity to such a small range of expressions. Shouldn't the watchmaker know how the watch is supposed to work? And if so, why are theses ideologies of inherent sin, and need for a intercessor, so harmful and even abusive?

Now obviously the argument for this is we messed things up when we fell from grace. But we've already concluded that this fall from grace was mankind becoming conscious and capable of judging between favorable and harmful conditions in our environment.

The point is, if your sacrifice was done for a reward then it's not a sacrifice, but an transaction. Love. True love, is out of compassion. And is not conditional. The beauty of love is it is not earned or fair. It doesn't require repayment because it isn't a transaction, but compassion and appreciation. It's not about making something acceptable. It's about accepting and appreciating what is, as it is!

"When you see your own pain staring back at you in the anger of your enemy you cease to hate your adversary."

Side note

(the early Christians were seen more as atheists in comparison to the other religious observances of Rome due to their abstaining form sacrificial practices. and thus they were subsequently blamed for all the calamities that befell Rome (similarly to the way Christian nationalists blame atheists for all that is wrong with the world today). The point is These ideas were radically progressive at the time but dangerously regressive this day in age.) originally the good new professed that humanity didn't need sacrifices to make them holy! And further more our value isn't tied to a religious identity. so you can see how these teachings and traditions have been grossly manipulated to preach a message that is entirely counter intuitive to its original meaning.

These traditions and stories most certainly hold profound wisdom and are not inherently bad. But they can and have been abused and exalted to a position where they are just as harmful as the restrictive practices and traditions they once opposed and sought to liberate humanity from.

Ultimately we may find that tho these categories of good and bad can be helpful in measuring degrees of things to which they may be harmful. It's important to recognize that the thing itself is not bad or good but rather the extent or context that dictates that.

Two examples may be fire and water, fire can be extremely destructive but when treated with respect it can be indispensably useful. And summarily water is vital for our survival. but too much water is just as detrimental as not enough. and so subsequently without our thirst, the water cannot quench.

Similarly The desire and drive are pivotal for our pleasure and prosperity.

By this point it should be apparent that the answer is not to completely reject our desires, nor is it to complete indulge them. But rather to listen and understand what they are telling us. discipline is not about repression and suppression but exercise and synchrissity. It's about growth and compassion. But this idea of a universal morality is not only problematic but counter intuitive because there will most certainly be potencies that are not healthy for everyone but necessary for some. And similarly there will be things that some people can't tolerate and yet others will be able to appreciate. And that's ok. That's the beauty of this world, it's big enough for all of us but far too vast for any one of us.

Culture and nature

In the book of genesis we are presented with this image of equilibrium in the form of a garden, where culture and nature cohabitate, as described by Jung in (a man in search for a soul).

It is in this garden where the fall from grace transpires, as we supplement gods prognosis of good! for our interjection of bad.

And It is here where we encounter the serpent, this creature that immediately attracts our fascination! because unlike ourselves and a great number of other creatures, this beast appears to be barren when it comes to appendages. It nether looks nor interacts with its environment like we do. And this alone makes this creature dangerous. but it is that very same quality that makes it useful. because it means we could actually learn something from it. However instead of being curious and developing the capacity to entertain a world much bigger than our limited view of it. We get hurt and close ourselves off thru the imposition of culture. Constructing a criteria and idea of what this world should be subsequently exiling ourselves form what is. And so when ever our nature would exhibit or express offensive or uncomfortable reactions to our environment, instead of learning to understand what our nature was trying to tell us about the very nature of nature. we imposed culture as a form of order to protect us from our natural proclivity for chaos. But the problem with that is much like nature, we too are not stationary objects. And that's why in regards to culture, life is not the absence of death but rather the art of dying gracefully. It's not in the order that strives to preserve it but in the chaos that tempts and teases it. That's the profound thing about love, because sometimes it's immaculate joy and others it's the deepest sorrow. But what ever shape it takes, it is all in. Ultimately it's saying no to everything else. so that it can give its everything to that one yes, it's the two worlds becoming one. And here arises the problem. because unlike nature that accommodates the full spectrum of life and death. cultures on the other hand are not so sympathetic to differing world views. This idea proposed by imposing culture to govern nature, looked more humane and fruitful (genesis 3:6). But as we soon found, this abuse of the "lesser or unfavorable entities of an ecosystem was not as beneficial as it first appeared. The ecosystem needs both the hot sun, and the inconvenient rains, in order to grow. (The accuser told us we weren't good enough and broken) so why have these traditions become more about degrading and judging then learning how to tolerate and appreciate life in all its shapes and forms?

Now I should clarify, anarchy is not the solution. And these abusive power structures have a vital purpose (in moderation). It was theses social structures that facilitated massive strides in science and technology. But it is often the rigidity of these systems that hinder and prohibit progress and growth.

My dear reader I can assure you that everything fades and dies. The goal is not to preserve it but rather to experience it so thoroughly, both the good and the (bad) that we appreciate the experience of what it was and don't idolize it as something it is not. In every shape life takes and in every season there is beauty as well as sorrow. We all have something profound to contribute to the ecosystem of life. But this idea that ether humans or ideologies have ether the capacity or the responsibility to fix someone els is harmful and demeaning. This world dose not need you to fix it. And you don't need anything in this world to fix or complete you.

Order is not the problem but nether is chaos. chaos is the raw material for creation and order the refined product of it. but creation isn't sustainable and order was secure but not fertile. The problem is chaos often looks like war because we are resistant to it. And order looks like an abuse of power because we idolize it. Both chaos and order are forms of life, and death, consecutively. Here's the point, nether chaos nor order are inherently good or bad. and further more that conflict between them is crucial for developing musculature. Chaos is destructive by very definition. And order is vulnerable to atrophy and corruption. But both are vital for life's cycles.

Our propensity for corruption and decay and our capacity for adaptation and growth are one in the same. With our exposure to potential danger we are also prevê to new discoveries, wonder and awe. And thus life is not simply the absence of death but rather the art of dying gracefully. It's not strictly in the order that preserves us but rather abundant in the chaos that threatens our categories.

Here's the point. There is no absolute evil, except for the evil of absolution. It is all good, in moderation. I'll say that one more time. Any good thing, in excess can and will become bad. But if you continue to see sin as some kind of cosmic debt incurred that then requires some form of transactional substitution to atone for. The the weight of that debt is a kind of hell. It results with us keeping score and ultimately being crushed by this incessant shame, fear, and enmity. Because in that paradigm the world and "god" are skewed against you and ultimately intend to cause you harm. The point is there is no cosmic debt. And further more if there was a debt and that debt was paid then there is no debt. Because a debt requires something to be owed. This idea that we owe Christ everything for paying our debt. Then infers that the debt wasn't payed, but was just transferred to another creditor. That we owed god and now we owe gods son instead. Ultimately the question remains. Is god good and loving, eager to bless their creation. Or is god a tyrant who creates creatures against their will, all so that they might monopolize on their creations misfortune. Or in other words is god like a loving mother who gives freely to her children or rather a malevolent businessman who bought up all the property so as to corner the market and ultimately enslave their patrons with outrageous prices?

Ultimately it comes down to what kind of world you believe in. Do you believe there is more than enough for everyone, a place where generosity is encouraged? Or do you believe in a small world with a scarcity of recourses, where the base line is defined by a massive debt far to large for the debtor to repay? A debt mind you far too large for the individual to incur but was incurred thru inheritance alone! Is this world one conducent for growth and discovery or one that cultivates stagnation and sterility?

One of compassion and curiosity or one of spite and fear?

Poem

Who is god to hate creation, was it creation or creator who chose it's being? Should creation be cursed for its curiosity, for its searching for god.

I do not say there is no god. but that divisive doctrine or contentious constitution are no good substitutions, to say that god is in this! but not that! is the most audacious heresy of all.

Who is God to make man in his image and then call man broken, and what loving mother would not answer when her child cry's out. Would they who made the heart and mind alike not speak all more fluently in the languages of both.

CHAPTER 6

APOCALYPSE

"The meaning of my existence is that life has addressed a question to me. Or, conversely, I myself am a question which is addressed to the world, and I must communicate my answer, for otherwise I am dependent upon the world's answer."
–(C.G. Jung)

One of the most useful tools our psyche gives us is the ability to sub out assumptions about our environment so as to allocate recourses and energy on more pressing maters. But far too often those assumptions are predicated on someone else's expectations. So not only do we find ourselves restricted by our own perception of reality but also those prognosis imposed by our culture. more often then not we can find ourselves basing our decisions on a map of the terrain that doesn't even fit our environment anymore.

Culture is a filter we use to quantify nature, but at the end of the day culture follows nature. Even as conscious as culture may claim to be, it can often be completely oblivious to its own nature and its relationship to the subconscious nature of nature itself. Because culture is an extension of nature.

This brings us to another common archetype which is that of the heretic.

In the biblical garden narrative we find this serpent which is described as cunning or crafty, (the Hebrew word here is "aroom") and it is the same word used later to describe Adam and Eve as naked or exposed. This same word is used elsewhere in the scriptures as wise, and it can easily be equated to the Greek word apocalypse.

Now if you'd recall, the reader of genesis was told that if Adam ate of the tree in question (the tree of the knowledge of good and evil) he would surely die. (I will point out here that the fruit in question is actually a fairly common image used in ancient mythology and most certainly in this context to convey the act of reading or comprehending wisdom. This is similar to the usage of the Hebrew word "haga" to study, devour, or consume knowledge.

This fruit is not to be taken as a literal fruit but as an choice to think rationally and critically. To take wisdom and knowledge for oneself and then use that information to judge the environment as good or bad (often solely with regards to how the environment serves the judge or individual in question).

And so tho a goat is killed to cover both Adam and Eve. I would argue that both Adam and Eve did in fact die when they ate of the tree. Tho perhaps not in the way we commonly understand death today.

In the book of Jonah we are told of a profit by the name of Jonah. And Jonah is told by god to go to a city named niniva. but instead he runs the other way. However as Jonah attempts to flea, his ship is stoped by a massive storm, and yet he is unmoved. Finally the sailors he is traveling with are convinced to throw him over bored. and here Jonah is swallowed up by a fish or a sea monster. Now we are told he spends three days in the belly of this chaos monster. But during that time he is referred to as being in the land of the dead or shaol.

I am not trying to convince my reader here that ether Adam and Eve or Jonah are in fact dead, as we understand it today. But where as WE define death as that moment when our hearts stop beating, for its original readers death had more to do with the reason your heart beat. Or being forgotten entirely. And so in the moment in which Adam and Eve ate of the fruit they ceased to exist as they had before. They were now concuss of times passing and abele to pass judgment on their suffering.

so back to this Greek word apocalypse. because as it is more commonly understood today it depicts the end of a world as we had known or understood it, and it carries a greater focus on the death of our illusions or perception of the world. However this word is more accurately defined as to reveal, uncover or make known.

Apocalypse = revelation.
Aroom = expose, make known

"For now we see only a reflection as in a mirror; then we shall see face to face. Now I know in part; then I shall know fully, even as I am fully known".
1 Corinthians 13:12 NIV

In the book of matthew we get this account of a woman at a well who recognizes Jesus as someone of great understanding. and so she proceeds to inquire of Jesus to resolve an ongoing dispute between the Jews and sumaratins at that time, as to where they should go to worship god. rather on the mountain or in the temple? To which Jesus replied "truly I tell you a time is coming and has already come when nether the mountain nor the temple shall be gods dwelling".

Nether the traditions of old(mountain). Nor the modern doctrine and dogma(temple).

But in humanity itself shall be sacred and holy.

We see this illustrated In this account in matthew 27:51 where The temple vail separating the holy of holies was torn in half. This event, rather literal or figuratively functions as a profound statement that god's presence and dwelling is no longer off limits to the common man and woman but that we are witness to the sacred all around us.

This revelation did not come without a cost however. For with this invitation to creation as a whole, we are required to die to our categories that divided us from it, or them,.

How do you desecrate the sacred?
You make the holy, commonplace!

The problem is without these stories and myths we often find our labors appear to be vain and futile. Or as Solomon put it "it is all meaningless or hevel" (Ecclesiastes 1:1-11) now this Hebrew word hevel is more accurately translated as mist or vapor. And it articulates not only the fleeting futility of our endeavors but the illusory nature of these myths that attempt to assert meaning to our toils and suffering. The point is not that our labors are in vain but that the meaning is not in exclusivity. In quantum physics there is a verifiable concept called quantum entanglement which describes the relationship between probable outcomes overlapping. Which asserts the interdependent nature of physics. What dose this mean? Well on the subatomic level when particles are in sync with each other they can interact with one another. However when these waves (because particles don't function as static objects but as frequencies of energy or waves) so when these waves are out of sync they don't interact. In isolation they virtually don't exist or lack coherent meaning. It's only in relation to everything else that physics exhibit's solidarity.

"energy cannot be destroyed or created but can change form" it's not that these out of sync particles cease to exist but rather that the state in which they exist cannot be observed by those occupying a different frequency. In this sense nothing really every dies but rather synchronizes with other frequencies.

And he died.

The profound claim of Jesus's death was that his justice didn't come in the form of gods wrath. But rather in the grace and compassion of the afflicted. Jesus didn't conquer the tyranny of Rome thru violence and might but rather thur unwavering resolve. This claim asserted that it wasn't wrathful power, or malevolent might, but compassion which was the defining trait of humanity. And further more there is no redemptive violence. As Jesus hung on the cross. He not only forgave is oppressors but wished them good. That they might see god.

The point was not divisive isolation.

In matthew 27:46 (NIV) Jesus cry's out to god saying "my god why have you forsaken me?"

This however was not a sign of defeat, but rather a victory cry. For you see Jesus is quoting psalm 22 here.

And yet Jesus dies!

In many places in scripture we are told that man cannot see the face of god and live. And yet Moses did?

The point is not that gods face is hidden from Jesus as he hung on the cross. But rather that he saw the face of god. It was the reality of who god was, that then incurred the death of our ideas or idols of god. And so instead of retaining his identity in his preconceived notions of what we thought god was, Jesus saw the face of who god truly is, and he died.

"for the river is not the same river, and the man not the same man".
Dowist dictum

Regardless of if you believe in the resurrection of the physical body of Christ, the physical embodiment of his metaphorical claim is profound.

In those days crucifixion was a political statement that the accused had no place in the Roman world. And with it brought shame not only on the culprit but on their families and anyone associated with them. However Jesus was not found guilty of any real crime. (this of corse is false because tho it was not one of violence, Jesus thru the title as the Christ or messiah did threaten insurrection and treason to the Roman authority). However it was not Rome but the religious establishment that found fault with Jesus as the Christ and pressed for prosecution. this act of submission to death on a cross, which was most certainly not an uncommon event by any means, and in fact there were many apocalypsists who died as martyrs around this time, but this was actually a statement that Jesus was also dying to any political or religious identities in the process. Which was in fact one of the claims the religious leaders had a problem with. Because Jesus ate with prostitutes and tax collectors, associating with the unclean and traitors. In fact this is a point Paul elaborates on later

"There is neither Jew nor Greek, slave nor free, male nor female, **for you are all one in Christ Jesus**.»
Galatians 3:28 (ESV)

But with this dilution of social, political, ethnic, religious, and gender identities also came the challenge of those authorities and denial of our allegiance to those authorities.

the sad thing is that the very man who sacrificed his life in opposition to the dogmatic idols of his day has himself become an idol. But that's nothing new, that's just how schools of thought work.

"The strength of a person's spirit would
then be measured by how much 'truth' he
could tolerate, or more precisely, to what
extent he needs to have it diluted.
disguised, sweetened, muted, falsified."
(Friedrich Nietzsche)

When questioned about paying taxes to Rome in mark. Jesus realizes he is ultimately being asked who's property is man? Dose he belong to Caesar or to God?

To which Jesus replies with another question "who's image is on the coin"
To which his accusers reply "Caesars".
Then Jesus said " give to Caesar what is Caesar's and to God what is god's".

For we bare gods likeness and tho we owe whatever debt we incur to our fellow man, as for allegiance we owe only to god and not to any idol. Which is what an identity is, an idol which claims to be our intended model or image.

But we are not stationary objects which are restricted to represent ether one or the other. But often these expectations or ideas of who we should be are imposed by our tribes and traditions, or our families and cultures. And so often the price for being one with the whole is abandoning our place of belonging in the exclusive part or division.

To call a stranger, brother and sister, is to denounce the exclusivity of your family's claim.

The story of the Bible from the very beginning is that we are all of one lineage, of one family. And we are all related, which genetically is accurate in how we as humans share approximately 99.9 percent of our genome . And roughly 50 percent with all life on earth. In fact genealogically we are not so distantly related to the stranger on the street.

We often forget this however because our concept of family is so small. it's solidity in the ancient world was even more so.

And so when Jesus rejects his family in luke 8:19-21.

He's not actually rejecting his family but rejecting the restrictive category of family. We see this again in Luke 14:26 (NIV) when he says

"If anyone comes to me and does not hate father and mother, wife and children, brothers and sisters—yes, even their own life—such a person cannot be my disciple."

As if the very price of inclusion with the whole is exclusion from the parts. we were not meant to be reduced or restricted to such a small identity. But to grow, adapt, and change. And from the very beginning both the messianic role and that of humans as the barer of gods image was to redefine what it means to be human, what it means to be alive, and what it means to be here and now!

This subversion of dualistic identities is not the problem but the solution.
We need a tradition or lineage to give us shape, and to give form to the fluidity of the whole. But the ultimate goal is to get back to that unified and complete state. We still need a structure to hold the thing together and obtain a degree of congruity and yet it's not the parts but the

relationship that permeates thru it all that retains the meaning. The point is to be reconciled with the whole and to move past those reductive traditions.

The first command god gives mankind is to be fruitful and multiply and yet the first thing we do is become divided. From Adam blaming eve and eve blaming the serpent, to Cains jealousy of his brother able who's gods favor was shown. And even this division illustrated thru the Tower of Babel. Where our own pride and desire to belong and be appreciated then drives us to reject those who don't look, think, talk, believe, or see like us and ours. And so it is no surprise when Jesus reiterates the greatest command is to love god and love others. Where as the way we do the first is to do the second, or be our brothers keeper.

Jesus expounds upon that in Matthew 5:43-48. When he commands us to love our enemies for this is the way we shall become perfect like god.

I don't think the point is that we should love our enemies despite our disagreements but rather admire and appreciate them because they see the world differently. It is because they see what we cannot, that then functions as a reminder that this world is not so very small but much much bigger than ourselves and our limited world view.

And how wonderful it is to live in a world where we don't have all the answers, where there is still room to discover and learn. A world with poets and mathematicians, conservatives who remember where we've been and progressives who are focused on where we are going, scientists and religious sages both looking for answers or god!

"the first gulp form the glass of natural science will turn you into an atheist, but at the bottom of the glass God is waiting for you"
Weren Heisenberg

Our humanity or our identity.

the profound claim behind Judas's betrayal wasn't that he betrayed Jesus (the human one) but that he betrayed his own humanity. Judas declared his allegiance to the religious and political authorities so as to attain admittance into the social hierarchy. but in doing so he betrayed his own nature. Which if you believe in a god, then you believe that god made us in their image. or in other words our very nature is that of god's. And yet all too often the church favors the agenda of a book over, the genuine nature of a human person when they condemn homosexuality, or interracial marriages (affirming racism). Our intuition is often vilified in many church's today because many religious and political leaders are more fond of the man made idols then actually knowing the god who's nature is observed in nature. For in the same way human nature is an extension of nature's nature. So too nature is an extension of its creator's nature. I know it sounds cleche telling my reader to be true to yourself. But This is more about being honest to yourself about what actually is, then it is about preserving our cultures egoic idol of what it should be. The truth is both ancient Egypt and the later Aztecs worshiped the sun. But nether of them actually worshiped the sun, they worship their personified ideal of the sun. It's only when we are honest about our own nature, that we are then able to know the truth about god and their nature. (now you may argue that the Bible is infallible and inerrant because it's words were breathed by god. But so were we! And you have no problem saying people are fallible and even inherently evil.)

Another good illustration of this can be observed in saul/Paul's conversion.

Because Paul was zealous for god and was bent on eradicating these heretics who professed blasphemy. And yet we watch this pious man become the very thing he hated in pursuit for god. Because at the time, for a Jew to become a Christian would have been like a Christian today becoming and atheist or agnostic. It was a complete denial and rejection of not only ones doctrine and traditions but one's identity. And it required denouncing the image or idea of who god was presumed to be. It meant leaving this image of god that was restricted by culture and orthodox interpretation. Rejecting a god of doctrine and monarchy, a nationalistic version of god that looks like us and not them (our enemies). And it meant opposing a god who's sense of morality had worked for us. but had abused or oppressed those who don't look and see like us. And so looking back on history it is often the heretic who is rejected by the small world and it's presiding deities for professing a bigger concept of god that is often revered.

"we look forward with fear but backward with distain".

Every tradition was once a revolution that opposed a restrictive tradition. These backwards and regressive traditions were once progressive yet palatable movements in the right direction.

In the 16th century the infallible authority on the shape and nature of the world was dictated by scripture. and so when Galileo challenged that protestation it was understandably a heretical claim. Because not only did he propose that the world was not flat but also that we were not the center of the universe. And for his claims he was denounced of any authority and prestige, subsequently shamed and shunned and even threatened with execution if he did not denounce his claim. Because the price of a bigger, deeper, world which was part of a much vaster universe was, death. It meant dying to our preconceived notions of how it all worked and where we fit or belonged. It meant loosing our small egocentric concept of the cosmos and subsequently our understanding of heavenly bodies as the heavens were god's divine dwelling.

The price of discovering something new often comes at the expense of loosing something we thought we knew. Perhaps the reason there are so many post apocalyptic iterations is because we crave the reinvigoration of chaos an deliverance from the atrophy of our current state of order.

As history often shows, those who speak for truth and god were labeled as heretics for challenging the fundamental tenants of there time.

There is no prolific wisdom which can equal the profundity of love. No amount of foresight or knowledge which can surmount the value of presence, of experiencing this moment, this expression of life, right here and now, as it is. There is profound beauty in the fact that people aren't perfect. Because Sterility is not sanctity. Life is chaos. And so It is truly good that we get to experience life, that we feel our emotions before our rational mind has time to reason such things away. There is an innate fluidity to human nature. And thus any creature you try to cage or hunger you neglect and starve, will inevitably become a monster. Not because this particular impulse or expression is evil. But because life's nature depends on expression. it is that vary nature which is contradicted by our attachment to a rigid identity. Matter can nether be created nor destroyed, just reallocated. It is This cyclical nature which is innate to all matter. And it highlights some of the pivotal clams of early Christianity.

at the root of this Christian movement we see passages like mark 3:30 and 1 Corinthians 7-9 Instructing converts to diss own their familial relations and even abstain from marriage. but what we forget is that these passage's were meant for people who were leaving their religions and communities. These people were abandoning their livelihoods and their place in the ancient world. But with this came This claim and assertion that our value as people is not dependent upon our role in a social hierarchy. For these people There were more pressing maters than retaining one's position in society. Matters more innate to our humanity than our subscription to a fixed world view. And so it is no surprise that these people who were abandoning their old world views for new ones were also instructed Not to take oaths (Matthew 5:38). At its heart this was apocalyptic literature. Yet it still adhered to its roots. We see in the book of mark this practice of baptism which is symbolic of Israel going thru the chaos waters (reed sea) in exodus which itself echoed the flood in genesis. This image is one of decreation or the deconstruction of order. Ultimately this practice of baptism is symbolic of a reversion to the original state of nature. And thus this is often described as both a death and a rebirth. However it's ties to the exodus go much deeper than the parting of the reed sea. Because this story was the anthem for practicing Jews. And so we see in mark, Jesus is baptized and the spirit of god descends upon him. At which point Jesus is led into the wilderness to be tested. This would have been a blatant claim that Jesus was a messianic figure as the one who looks like god and is chosen to go on god's behalf just as mosha had. but Jesus's name wasn't Moses but rather Joshua. Now to the modern English speaking reader this may sound absurd, Jesus's name was Jesus not Joshua, however in Greek the name Jesus was an abbreviated transliteration of the name Joshua or perhaps more accurately yusha. Which is more fitting because Jesus didn't deliver Israel from the abusive tyranny of Rome but rather lead Israel into the promised land bringing about gods kingdom on earth. which was the role Joshua undertook. The claim here was that god didn't look like the tyrannical Caesar or malevolent master but rather the compassionate human in tune with nature. god was not the abusive power structure but the loving voice of curiosity. And so tho these passages offer instruction, they also recognize something far more profound. Which is resonance with life. This dose not offer a destination but rather simply a direction. John reiterates this much later by acknowledging both the end of the world as they knew it as well as the fruition of a new heaven and new earth. Or in other words. Our view of the cosmos would soon change.

CHAPTER 7

THE KINGDOM OF HEAVEN

There is a phenomenon commonly know as Stockholm syndrome where a victim becomes emotionally attached to their captor.

Similarly We can become attached to our trauma (which here simply refers to unresolved experiences). Freedom is frightening and bondage can provide a sense of security, further more pain and fear can be effective motivators. But what happens when we resolve that trauma and find healthier habits to replace the harmful coping mechanisms? What do we do when we lose all the fear and shame/pain which motivated us? What happens After we have abandoned those dualistic paradigms and mindsets of scarcity? Is there life after death? Can we believe in an resurrection? Or dose the death of the disease result in sterility? Devoid of both corruption and life respectfully! Thus far in our journey We have left the abusive order of Egypt! We have deconstructed these doctrines and traditions in the chaotic wilderness! And now it's time to enter the land of promise. The question is how do we live in the destination when our whole lives up to this point have been about the journey?

"And those who were seen dancing were thought to be insane by those who couldn't hear the music"

–Fredrick neiztchie

"The transcendently beautiful thing about grief and trepidation is they do not require validation but simply demand to be witnessed, almost as if to say in some magnanimous way " I am bigger than you, and if you simply listen to me, I intend to make you bigger too!"

–M.R. Holt

There is a direct correlation between our deepest fears and our deepest desires.

For instance, say your deepest desire is to be accepted by a person or a group of people. Your deepest fear then would probably be rejection. and yet the ultimate goal is to be fully known by those people.

Now say your deepest desire was to achieve success, subsequently your greatest fear then would be failure. And yet the ultimate goal is a finished product.

If you were an explorer looking to discover a new land, your greatest fear might be getting lost along the way. However the ultimate destination is one where no one has ever been before.

As often was the case for those heroes of old. They were only immortalized after they had sacrificed their life to a cause bigger then themselves. Ultimately, rather it be the fear of rejection or failure, it is only after we have confronted our greatest fears that we are permitted to witness the fruition of our deepest desires.

There's An interesting point here, because Christianity is predicated on the presumption that gentiles understood the Jewish scriptures and messianic role better than Jews. which would be like saying atheists have a better grasp of the prolific messages in the Bible then the devout Christians who idolize the texts today.

As we have already stated. It is often the heretic who challenges the orthodox traditions of their day, the one who deliberately wrestles with god. Who then finds god and truth in the process.

Heaven on earth

In the New Testament we get a collection of stories describing the kingdom of god. in matthew 13. Jesus begins with a parable accounting a man scattering seeds on the ground. And He elaborates on this story by explaining that the seeds are the "good news" (a bit of Greek propaganda used to announce the birth of a Caesar or coming of an kingdom). Here Jesus states that some hearts would be hard and would not accept the news and yet others shallow and fleeting, and yet even others which would misinterpret the message, all of which would wither up and die in time. But a few would take the message to heart, they would wrestle and nurture it, and it would reap 160 times what was sown. He expounds upon this by claiming that they who live with a mindset of scarcity will find themselves in such a world. but to them who believe in an abundant world with more than enough for everyone, they shall see that world come to fruition.

In verse 34 of Luke he continues, but this time with a collection describing the kingdom as a precious treasure hidden in an field. We are told that a man finds this pearl and sells all that he has to obtain the treasure. it is clear here that the price for the treasure is worth more then everything he has or knows.

Because that is the price for a land of promise or a world of abundance.

It requires us to leave behind our limited world view and scarcity mindset.

In the very same chapter of Luke Jesus tells a story of a man who throws a massive feast, but when the time comes for the party to begin, all those invited are distracted or busy. And so much like the muse discussed in chapter 3 of this book, there will be those who are preoccupied with the noise and end up neglecting the music entirely.

In matthew 19:13 small children are kept from seeing Jesus. But when Jesse sees this He rebukes his followers and declares.

"Let the little children come to me, and do not hinder them, for the kingdom of heaven belongs to such as these." Matthew 19:14 (NIV)

For children are curious and eager to lean and discover. They listen and fixate on the simplest things. Children are naturally intuitive and do not act out of a sense of duty but out of an honest desire to understand and know.

They are not bogged down with all the expectations or exemptions but rather driven by an incessant muse or vivid imagination.

Take for instance this account in John where Jesus divulges the way and means in which the kingdom of god may be obtained.

"Very truly I tell you, no one can see the kingdom of God unless they are born again." John 3:3 (NIV)

The psychologist Jordan Peterson has a quote from one of his lectures that articulates this perfectly. He says and I quote

"you have to decide which you love more, what you know or what you don't." (Jordan Peterson)

Because what you know provides a sense of security and stability but it also becomes stagnant and sterile. Where as what you don't know can be very offensive and even frightening. But it also is the only means in which we grow. life is the process of continually dying to what we know or think we know, so that we may grow and ultimately live!

Because sterility may be devoid of corruption but it's also devoid of any life.

The second Adam (human one)

In Luke 15 Jesus tells a series of stories describing what the son of man or the human one looks like. And ultimately he concludes by summarizing what it means to be human, quite simply as being your brothers/ sisters keeper. But he dose this thru the employment of 3 stories. The first of which is about a Shepard who looses a sheep, and so he leaves the 99 in search of the one.

In the second story it is a woman who looses a coin and searches tirelessly until she finds it. Now we are told that when she dose find the lost coin she invites her friends over to celebrate with her.

But the third story is slightly different. Because the third story is about a father who has two sons. one older and more traditional son and one younger more unconventional son. Now in this story the younger brother ask's his father for his inheritance, and this would have been essentially like saying to his father I wish you were dead or as far as I'm concerned you are. Basically he's saying all I want from you is your wealth or what you can do for me. But as audacious as this request was the father gives him what he asked for and before long the younger brother leaves ultimately squandering his inheritance until he is lower than the swine.

(an interesting detail is this word for property of which the father divides among his sons, in the greek this word is "bion" (bio) is the word for life.)

Now after a wile the younger brother humbly returns to his father to work as a servant in the hopes of repaying his portion of the inheritance and buying his way back into the family. but at seeing his son still a long way off, the father kills the fattened calf and throws a feast in his honor because his son was dead and is alive again. But not everyone was happy to see the son return for when the older son/brother returned from working in the field and heard the music he was furious and refuses to join the banquet. the faithful brother was bitter. and honestly can you blame him. He had worked much longer for his father trying to earn his place in the family and so it's understandable why he would feel betrayed. Now when the father sees his son refuse to join the party, he goes to his son and explains to him that he had always been with him and everything The father has was his now. which was right because the younger brother had squandered his portion of the inheritance and could only rejoin the family at his brothers expense.

But you see both sons had missed the point. For the younger son went looking for belonging, where as the older son tried to earn his place in the family. When the truth of the mater was they both had belonged the whole time. And tho for the younger brother it was shame and guilt that separated him, for the older son it was pride and duty that kept him from the father. Now there's an important detail, because in its context it would have been clear that the true older brother would have gone looking for the one who was lost. And yet I look at the church today and I see leaders more concerned with being right and criticizing those who are lead away (deconstructing their religion) rather then actually hearing what they have to say. The church is more concerned with preserving its myth then it is with allowing these stories to grow and evolve with people as intended. and personally I remember for myself it was this attachment of my identity to an ideology that kept me from loving my enemy for seeing what I could not. and it was ultimately rooted in fear of becoming lost too. The truth is I don't know what comes after death, if we just cease to exist or if there's an heaven or hell in the next life. but I do know there's a heaven and hell in this one! because that's where we find the older brother in the story consumed by hatred and spite and this incessant need to be right! and it's no better than where we find the younger brother riddled with shame and guilt. when it is apparent that what both truly desire is a place to belong and to be accepted as they are.

Getting out of the boat

In the book of John, after Jesus was crucified he appears to 6 of his followers near the lake of Tiberias.

After suffering a devastating defeat at the loss of his rabbi and denying even knowing his teacher 3 times, petter returns to his old life as a fisherman doing the one thing he was good at. However he appears to have lost his touch as a fisherman, for we are told Peter stays out on the lake all night and catches nothing.

And so Peter and his companions head back to the shore with the breaking dawn. when all of a sudden they see a strange figure on the shoreline who calls out to them saying " throw your nets on the right side of the boat".

They obey, still unaware of who this strange character is. And as they cast their nets, they immediately catch copious amounts of fish.

At recognizing the stranger on the shoreline to be his rabbi, Peter jumps out of the boat and begins swimming to jesus.

Now before we continue we need some context. Because we have a similar story regarding Peter jumping out of a boat in pursuit of his rabbi. The fist account of this phenomenon is found in Matthew chapter 14:24-33. In this account however Jesus is not on the shore but rather seen walking on the water. Here too the disciples did not recognize Jesus as their rabbi but instead hypothesized the figure to be a specter or ghost. Upon seeing his disciples in fear, Jesus calls out to them instructing them to take courage for it was I (Jesus) their rabbi coming towards them.

At hearing this, Peter test Jesus by saying, " if you are in fact my rabbi, then tell me to follow you"

And just as Peter had asked, Jesus calls Peter to him.

At heeding his rabbis call, Peter got down out of the boat and began to walk to Jesus on the water.

However before too long Peter began to doubt and found himself sinking.

It is here where Jesus reached out and grabs Peter saying " you of little faith, why do you doubt"?

Now it's clear here that the one Peter doubted wasn't Jesus. Otherwise he wouldn't have cried out to his rabbi for help! The person Peter doubted was himself.

And so when we see Peter again in John 21, Peter isn't concerned with staying above the water, with success or failure. Because he had already been defeated. Peters only concern was following is rabbi or teacher.

He didn't have to test god to make sure he heard correctly.

Now a key detail about Jesus after the resurrection is in a good number of instance, those who would have known Jesus best did not recognize him.

And I always found It odd. but they weren't expecting jesus, or in other words Jesus wasn't what they were expecting. This wasn't what they had in mind when they thought of the son of god or the Christ. And how often is the same true for us. Because we have this idea or cosmological map of what it's supposed to look like.

And we see this idea depicted in the form of an ark. This vessel we use to navigate and interpret the seemingly chaotic nature of life, these cosmological chaos waters of decreation. And in this ark we order the cosmos, constructing a world we can comprehend, putting two of each creature, explaining how and why things work.

But we see this ark all throughout scripture.

We see it in exodus as the vessel which delivers mosha thru the waters and from death at the hand of Egypt. And again as an ark of the covenant. And where ever this ark goes, there too is where god is. Thus far we've painted a rather abstract image of this ark. But in modern terms this ark could be described as a doctrine of ideology. This map of the world and how it supposedly works. And so in the first case this was god giving mankind rest. Delivering man form chaos thru the intercession of order. How ever on the exodus account this was god delivering mankind form a supreme utilitarianistic form of order and into a chaotic wilderness where they could deconstruct these useful but abusive and oppressive ideas. Delivering Israel from a scarcity mindset so they could get to a healthy abundant one, A land of promise if you will.

But as we see at the foot of the mountain in exodus 19, chaos is frightening and the ineffable more so. So god gives them the law, the 10 declarations. And this is not only god teaching these people how to treat them selves and others as holy, but also gods way of holding them closer and tighter.

This law and the representation of the ark was gods way of giving Israel a sense of security and stability in a chaotic new world.

And so when we see Peter get out of the boat and walk to Jesus or even later jump out and swim to him. We recognize the imagery here. Peter is leaving his map of the world, his conception of how the world works, these doctrines, dogmas, traditions, and ideologies of who he thought god was, all to rediscover who god is in pursuit of the Christ or the one who looks like god. Peter knew the boat, he was a fisher man, this was who he was, this was his identity as a Jew, this was his utility, how he fit into the world, the role he filled in the Roman Empire. And he left it behind for the unknown.

And this is the repeating story of the Bible not to mention human history. From Abraham to Galileo. It's us leaving these maps of our terrain, surrendering these images of who we are, and these ideas of what we think god is, all to rediscover a whole new world.

What it means to be human is to recognize a god that doesn't fit our expectations, to leave behind these expectations of what we thought god was, to rediscover what god is here and now. What it means to be human, at least according to jesus, is to recognize we are all connected. And this is love, to see yourself in creation and creation in yourself.

Side note

(there is virtually no extra biblical evidence to validate the historicity of Moses or the exile from Egypt) In Hebrew the word Moses means to "draw out", but in Egyptian it means "son of".(also in Hebrew the word bal means "master")

Life after death

"Our god is The god of the living not the dead" Jesus (Matthew 22:32) NIV

"The living know that they will die, but the dead know nothing. For they are forgotten." Solomon (Ecclesiastes 9:5) NIV

Buddhists believe in reincarnation, that is after we die, we come back as something els. This is where we get phrases like, you were a - - - in a past life. It is also used as an explanation for deja vu. tho this is probably a result of our genes communicating ancestral information to our psyche. However There is probably some truth to this idea of reincarnation. Because as our body decays, the cells are absorbed into the earth and then repurposed for new life. And the conditions in which that life takes root will be a result of the world we left for it. And so in some degree if we leave the world better off then we found it. Then the environment that those repurposed cells inhabit will be a direct result of that better world. And similarly, tho our offspring won't be reincarnations of us, they will carry on our genetic code packed to the brim with information we helped create.

"A society grows great when old men plant trees whose shade they know they shall never sit in."

Seeing as how this chapter is entitled the kingdom of heaven, I should probably address a fairly common question, which is what comes after death? But to be perfectly honest with my reader, I myself have never died. That is in the literal physical sense. I do however know what it's like to loose a loved one. To watch someone you cherish pass away. I know that in time our bodies decay and our memory fades. In fact I also know that those experiences influence latter generations and those decaying corpses cultivate the earth. I know for a fact that there is life after death, that is death is a crucial part of life. I do not know however if Jung was right when he hypothesized that our conscious state persists in some form of psychic unity. Or if there is in fact a place where our conscious awareness will spend an indeterminate amount of time. I don't know if death is essentially the same as being born or if there is a ethereal heaven and hell. I do know however that there is both In the here and now.

I have seen people and even myself at times stuck in a perpetual anguish and torment as a result of our distain, resentment, bitterness, hatred, shame and spite. I've seen harmful responses to fear and grief absolutely destroy people. I've seen intolerance and prejudice isolate entire groups of people. There is a hell! But more often then not that very hell is the result of religion.

But there is also a heaven. A mindful presence (sabbath) an open acceptance and appreciation, an eternal state of compassion and love. There is hope and healing and an abundant life. The truth is we are all connected and so we are only ever as alone as we believe we are. We can be content in solitude, and we can be restlessly isolated in a crowded room. But ether way there is a death required. Ether we die to our dualistic view of a world divided against itself or we die to the one that is connected and unified. Ether we kill the ground in an attempt to keep it holy, or we dissolve our divisions of the whole.

Don't get me wrong. I don't want the church to die, I'm not trying to kill god. I want revival. But the nationalistic, dogmatic idols. Those need to go. Our rigged suffocating traditions and rituals need to be allowed to grow and evolve. In order for there to be life, these once good strides in the right direction must die. Because every tradition was at one time a revolutionary new idea. Heaven on earth is realizing that the very ground you stand on is holy. That all life is connected and a crucial part of the whole.

There is a Buddhist practice where the individual will meditate with the soul design of rejecting oneself and their individuality from the universe in the process. This practice seeks a state known as zen where the observer finds resonance with life itself by denouncing the ego.

And tho I myself personally disagree with the conclusion of this practice (which is that their is no self but that we are the universe whitening itself) I do however find profound insight in both the rejection of duality and the acknowledgment of the universality of life. " who is Calvin to denounce creations inclusive claims in favor of man's restrictive dogmas"?

the truth of the mater is no one on this earth had any say in their existence and so the idea then that we will be judged for all of eternity in some cosmic heaven or hell based on a standard that designates us as inherently and irreparably evil for simply existing and then condemns us not based on our capacity for love and compassion but rather based on how thoroughly we adhere to a man made doctrine. Is not only absolutely ridiculous but more so extremely harmful to our mental health.

I mean dose that conception of an heaven and hell really make any sense? Dose it sound just or good? And if it was the case then the best most compassionate alternative would be to cease procreation entirely. And yet the very first command in scripture is "be fruitful and multiply". The idea that according to this world view, good, kind, and compassionate people will suffer for all of eternity all because they developed critical thinking skills and spoke out agents harmful systems of oppression, all wile hateful, spiteful, and ignorant people will be rewarded all because they were born to the right people at the right time and they never questioned injustices perpetuated by the systems they subscribed to. Not only will the compassionate be judged for their love, but the un-compassionate will be rewarded for their lack of empathy. And all of this cosmic justice follows suit with the man made doctrines!

"If the Bible says it, I believe it, that settles it".

My reader may ask, who am I to question scripture! I am human! and that is in fact our role. Especially when these ideas and practices are hurting people. You see the Bible it's self is not only riddled with contradictions but actually deliberately contends with itself. For example

Jeremiah diss agrees with Ezekiel in regards to precepts in exodus commanding Israel to sacrifice their first born children to god. (Exodus 22:28-29)NIV (exodus 13:2) NIV (Ezekiel 20:25-26) NLV and (Jeremiah 7:31; 19:5; 32:35) NIV where exodus requires on behalf of god, that all first born sons should be given to god. Ezekiel condemns this request as barbaric, and Jeremiah goes a step further and says that god never required it at all. You see the Bible was not written by god, but by man. And further more it is not a univocal book but rather a collection of texts that account the progression of mankind form primitive tribal people to a civilized society. And it is most certainly our responsibility to rectify these ideologies that no longer server their purpose but instead become harmful to humanity.

It is my solemn belief that every single person on this planet is trying to love, themselves and others as best as they know how. The problem is we show the love we believe we ourselves need. And so if we crave stability and security the love we show can often be restrictive and rigged.

If we feel broken and don't know how to handle our hurts, our love can often manifest in ways that treat people as problems or broken things that need to be solved or fixed.

But we look at Jesus and how he loved. And tho often that looked like healing the sick and feeding the hungry. sometimes that looked like saying no. and even others it was simply sitting shiva, grieving with those who had lost loved ones. Here's the point love doesn't have to have all the answers. Sometimes love and even life are in the pain. There's a Dowist quote that states

"we make room within ourselves for the immensities of the world".

Often it's not about fixing a problem but simply learning how to appreciate and love what is, even if that looks like grieving what was.

Before we conclude I would like to thank my reader for listening to these words. And reiterate that you dear reader, are whole, complete, and holy. In whatever shape you take in this moment here and now.

Conclusion

All too few actually feel the rain on in their souls, the rest simply get wet.

(roger miller)

(If you have made it thus far in the book I have probably offended you on more than one occasion, and yet you are still reading! Good! That means you have contended with something uncomfortable, an idea that prior to the reading of this book you had lacked the capacity to tolerate. And further more it means that you have grown. Congratulations on stepping into your humanity!) because that's the whole point;)

my reader might understandably presume that I the author sorely loath or hate Christianity. This however is not the case! I draw from its stories and traditions because I still find profound wisdom in both. And tho there are most certainly forms of Christianity that I would eagerly encourage to die. It is only so that these stories and traditions might be allowed to truly live as they evolve and grow as opposed to submitting themselves and those who subscribe to them to a form of atrophy in the confines of an echo chamber all in the name of posterity. The profound thing about these stories is that the Bible is not a univocal book but rather a diverse collection of discussions and dialogue each approaching the experience of existence from an broad array of perspectives.

The nature of life

If you haven't already caught on. Life by very nature is impermanent.

Ecclesiastes 3:1-3 KJV
For everything there is a season, and a time for every purpose under heaven: **2** a time to be born, and a time to die; a time to plant, and a time to pluck up that which is planted; **3** a time to kill, and a time to heal; a time to break down, and a time to build up;"

For you see life is everything it is, until it is something els entirely. And we are no different. By that I mean we are all quite different indeed.

So what ever you are right now. What ever you are working on in this moment here and now. It is both the most precious thing you will ever have and yet at the same time fleeting and fragile.

There will be times when you are a brother or a sister, times when you are a father or a mother. Times when you will be a spouse or an friend. And these people are beautiful roles to fill. But there will very likely be times in your life when you are none of these people. And that too is ok. We are not the roles we fill. And listen when I say this because you are whole and enough on your own! I repeat, you do not need someone else to complete you! There was a time when we needed our place in this world to be validated by a title or role. But our identities can't be restricted to our relation to our communities. Because these titles and roles are only a small part of what it means to be you!

Do you remember that time I told you, your were enough on your own. Yeah I know it was like two seconds ago. I mean every word of it. But that doesn't mean we then get to isolate ourselves from the world. The point of abandoning these categories is to recognize our role in all of life's shapes and forms. You see because, what it means to be human has everything to do with the ineffable. It's about wrestling with god, contending with the conflict. And so those things that are bigger than you! The ones that are uncomfortable, frightening, and even painful. I would encourage you not to steer away from them. Because tho you most certain are enough as you are. The truth is you are so much more then that.

And if you take anything away from this book, I hope that it is that no mater what you are. It's never too later or too early to rediscover and redefine what it means to be you, here and now. And that's not mysticism, that's just the nature of life.

But what if it's not good enough? What if you give your all and receive no glory in return? What if you take your big chance and fail? That's part of it. The true cost isn't the energy and time but the expectations we have for our creations. And in order for anything to truly come to fruition, we have sacrifice it's potential to be more then what it is! it's not our responsibility to ensure that our endeavors succeed, our only job is to create what we are here to make to the best of our abilities. We are here to do and say what we've been given to do or say. Maybe your revolutionary idea isn't the one that brings about change, but rather it is just a contributing factor. Never the less, it's a step in the right direction. Perhaps this life isn't a race to the end but rather a dance. It's not about moving forward or backward, but stepping gracefully and being in sync with the music.

Among the 10 commandments given to Gods people was the command not to take the lords name in vain or more accurately not to bare gods name in vain, but to do all things as if in service to god. And so as is still common today, people would swear oaths or make pledges with god as their witness. and so in Matthew when Jesus instructs people to refrain from making oaths, that is

"let your Yes be yes and no be no for anything more comes calamity"
(Matthew 5:37). New KJV

He's addressing the futility of life, for such certainties we have little to no control over. The vow is us isolating ourselves from the whole and resigning our expression of humans to an identity or part.

Gamaliel may have put it best in acts when in regards to the Christian movement, he argued on behalf of the Sanhedrin's actions and retaliation, to do nothing.

"For if their purpose or activity is of human origin, it will fail.

But if it is from God, you will not be able to stop these men; you will only find yourselves fighting against God."
(Gamaliel) (Acts 5:34-40) NIV

Essentially that if it's meant to be it will be, and if it's not, then it won't be.

How to approach life

"The maturity of man, that is has reacquired the seriousness one had as a child at play"
(Fredrick neiztchie)

Basically take it playfully not parlously, be more curious then cautious.

It's serous and your life depends on it. But it's also completely out of your control. This is your life, here and now, in this shape and form. And yet it's also temporary, it's everything it is, until it's something else. but if your hung up on what it's not, then you'll miss what it is.

The truth is we don't really want happiness but rather the pursuit of happiness. It's the way our bodies are made. To crave what we need until we don't need it anymore. without the hunger there is no pleasure.

Suffering is how our bodies tell us something isn't right or that we are lacking a vital necessity. But in many cases it simply communicates a loss of something we had, warmth, companionship, validation, sustenance…. And tho there is great pleasure when these needs are met in excess, there is very seldom much recognition emphasized when what we've come to expect is met. we want what we can't have. And this is the power behind flirting. (Giving someone a taste and inspiring an hunger that was previously absent or unaware). pleasure without pain is monotonous. happiness is not a destination it's a direction, it's a goal we pursue not an objective we obtain.

And this is a good thing because it fuels our curiosity. But there's a flip side to this. Our drives and desires are a crucial part of what it means to be human. But So is our contempt. A word which often carries negative connotation but in this sense bares a positive one. Because sometimes we are what we lack and sometimes it is that lack that makes us what we are.

The precious memories and moments in our life are like fresh fruit from a tree. They are sweet and good but not permanent. They are meant to be enjoyed, but not kept. Eventually you will have to let go. And let that good thing die. Because otherwise that fresh fruit, that precious moment, will rot in your mouth and you will become bitter. Much like these precious moments or this good fruit, we have preserved traditions which hold profound wisdom and prolific insight. But also like the fruit or the memory, if we don't let these traditions grow, evolve, and even die, they too will become stagnant and sterile. absent of mutation but also devoid of any life. The goal of sanctity cannot be isolation or intolerance. Similarly our hungers may be quenched but should never and truthfully can never be completely mitigated. Our genes prerogatives have a purpose of paramount importance. And the prerogatives produce far more than preservation and survival. But procreate new life.

The good news.

Truly I tell you, some who are standing here will not taste death before they see the Son of Man coming in his kingdom."
Matthew 16:28 NIV

From Jesus's own mouth comes some of the most compelling evidence against the Christian faith. And perhaps that's exactly what it is. because ether Jesus was wrong and worse a lier, or eternity is at hand here and now in our mindful presence and appreciation. In ether case we should not be waiting for Jesus's return like some clandestine deliverance form our decrepit state, but paying heed to god's presence. appreciating all ground as holy, all moments as beautiful, and all expressions or experiences as good and crucial. There is a story in the New Testament where Jesus attends the funeral of a dear friend named Lazarus. And tho Jesus was going to raise Lazarus from the dead. His first response is not to fix the problem but to be present in the grief with those morning the loss of their loved one. Here Jesus simply weeps.

Because you see, people don't want you to solve them like problems but to love them in this shape and form, as they are. I'll say that again. we want to be seen and appreciated, not fixed and resolved, happy, sad, tired, anxious.. these are all states we inhabit and expressions of what it means to be us. And Jesus recognized this expression of grief and sorrow as a beautiful and profound testament to life.

Do not mistake me, we are not our sorrow nor are we our elation, but they too are vessels for our conscious minds and intuitive hearts to find common ground.

For Love is not to say this I am or this I should be, nor is it this you are or thus should you be to be worthy of belonging to me. Love is not an expectation but acceptance, not complacency but compassion, not one thing to many but many to one. Not a problem to be solved or reward to be won but presence to be admired and even acknowledged for its being.

We all want to be seen.

But it's not until we learn to see the unseen thing within ourselves, that we are then able to truly see the world, and ourselves in it. true real and honest love is not about fitting our expectations of what (should be) but appreciating what is.

The will of man and god

From Zoroastrianism's predestination to Jung's prewired instincts, the free will of man has seen it's challenges. However it is nether the preconceived notions of doctoral gods or those subconscious impulses of our psyche that dictate our corse of action. Ultimately we chose how to react.

It was our fall from grace (our becoming conscious of our instincts and questioning our existence) that then separated us from nature and yet on an existential level United us more intimately with it. This prolific event then permitted man to contradict his instincts and assert his autonomy over nature's algorithms. We get to rewire our neuro pathways, we alone decide which heavenly voices assignment of meaning and value to adhere to.

Tho I will admit It is a delusion for us to believe that our ideas belong to us. When the truth of the mater is, we belong to them.

The muse chooses the artist, not the other way around. The only choice we are left with is, are we going to answer the call. Are we going to listen to what the world has to say, and have the courage to follow it (and at times challenge it). Or will we plug our ears in willful ignorance for god's heretical claims!

Because the cool thing about the muse, is she can't be caught. He can not be pined down or tamed, they can't be stopped. No the muse will flow. I still believe things happen for a reason and that life is building up to something profound. I believe there are prolific messages in what the world would call trivial and mundane. And I still believe in god, at times anyway! But we've all got a story to tell, and we are all part of the big picture in the end. But that picture is far to big for any one of us. And so we get too experience that picture the only way we can. In moments!

The title Jesus alluded to most was the son of man or the human one, essentially making this claim that what it means to be human ultimately has nothing to do with our religious and political affiliations and further more those corresponding identities must dissolve in order for us to truly embody our humanity. This is exemplified in the way Jesus was deliberately rejected and executed under the guise of the religious and political authorities of his day.

"For unless you die and are born again you cannot enter the kingdom of heaven".

For Jesus, our humanity has far more to do with how we treat people then it did with what stories we subscribed to. And further more the kingdom of god or heaven was more about being connected to life here and now then it was about going to the right place when we die. It was no mistake that virtually all sentient life forms on earth express a proclivity for curiosity, we are innately prone to oooing and aweing over things. We have this incessant desire to admire and adore. And yes we want to be appreciated ourselves. We want to be seen! But it's not until we learn to see the unseen parts of ourselves, that we then truly see the world.

For we are part of that world. And love is seeing ourselves in and thru all the divers expressions of life. Or as Jesus put it "whatever you have done for the least of these you have done also unto the son of god" or perhaps even god themselves. For to truly love god is to love gods creations. To recognize the ground on which you stand as holy and to see the trivial and commonplace as prolific and profound.

Because you see, from Abraham to Paul, and everyone in between. this deconstruction of religious and political identities is what the whole story has been about all along.

If you believe in a source from wince all creation came then to some degree you acknowledge that all life in some form or fashion resembles that source or origin by extension.

Racism

As a final remark before concluding this literary work. I will address a topic I have as of yet discussed explicitly. And that would be the subject of racism. For nether in its ancient religious sense nor in its more modern Darwinistic context do I condone any perception of mankind that then degrades or dehumanizes a group of people based on ether ethnic, or cultural traits. Nor do I defend biblical passages that lack rebuke for any forms of slavery that propagates the idea that people are objects or property meant to serve systems that were intended to serve man. Any claim that an individual of a different race, gender, or class is lesser in any fashion, is inexcusably repugnant and will not be tolerated in any form in this conclusion. However this particular fight is not my story to tell. For we all have a story to tell and to take one that dose not belong to me especially when I have so much that is mine to discuss would be wrong. And so that is all I shall interject on that particular subject.

The answer to our question!

And so what it means to be human and what that has to do with the ineffable, is not the question but the solution. Because what it means to be human, is to be both nothing and everything all at once. To be heaven on earth, the transcendent attempt at empiricism. But above all it is to be this moment in this form. It's abstract and ambiguous, but that's far better then being restricted to an identity. What it means to be human is to wrestle with god, to contend with the ineffable, and to entertain the abstract, ambiguous, and chaotic qualities that can't be quantified or categorized. Because that abstract chaos is what connects it all together. It's the gravitational force and interconnected relationship between all the crucially integral parts that gives the experience its meaning.

It is only thru contending with the ineffable that we then develop the musculature and capacity to tolerate and appreciate a GOD that is, as opposed to the idol we expected. Life's purpose is not to have as much pleasure as possible. But to find fulfillment and connection. It's goal is a deeper compassion with people and experiences that look and act differently. It's about finding resonance with the full spectrum of existence, in all its dichotomy. Because By very nature, life's primary prerogative is to discover and grow, to adapt and redefine itself "life". To take new shapes and embody different states of being. God is still inviting us onto the holy mountain with them. To be a nation of priests, to be to divine dwelling place of god. To rediscover and redefine what it means to be human. In all the joy sorrow and anger (because they are not mutually exclusive) there is dichotomy and diversity, the duality is the anomaly.

And so what does it mean to be human, what dose it mean to be you? This is no question for lofty sages or sacred text but for the one asking the question. Who are you? Well who's asking?

in the words of Carl Jung "I'd rather be whole than good" I'd rather have the capacity to appreciate all of life's expressions, then be restricted to but a small fraction of it.

nature is not broken and in need of fixing. It's the abusive human systems which have exploited life that are the problem and in need of resolution. It's not our instincts and intuitions but our doctrine And dogma that need to be rectified. The solution is the god who created nature and not the god created by culture. The god who invited Israel, as those who contend with the ineffable, to be god's partners. Not the traditions which had become an idol, the bal which had become an malevolent master. But The god who delivers humanity form the abusive power of a supreme utilitarian order and into a chaotic wilderness to deconstruct the abusive ideologies. To abandon the mindset of scarcity and enter into a land of abundance and promise. I believe in the The god who called Abraham to leave his tribe and traditions and Saul to become the very heretic he once hated. I believe in the god who taught us to treat the very ground we stand on as holy and who's authority is not in some sacred name but in presence itself "I am". My heart belongs to the god who sought the stubborn mule as their companion and professed the foolish stork to be their dearly beloved. The god who called chaos herself their crony and kindred spirit. (Job 38-42) the god who dose not require a sacrifice to atone for our existence because people are not problems to be solved and life is not a broken thing that needs to be fixed. But the god who looked at all they had made and called it good, even making it their dwelling (sabbath). This idea that god cares more about our subscription to an ideology, And An attachment of our identity to a doctrine then they do about how we treat others is completely counter intuitive. How did a message that professed we were the bearers of gods image and that we didn't need a temple or holy mountain to commune with god. Become more concerned with the divine authority of scripture then our own humanity. How did a god who called us to love our enemies as invitations to a bigger deeper understanding of the world we live in. become the dualistic god who only loves those who look and think like us? When did the god who created the universe become restricted to one cultures interpretation? I honestly don't care how mighty and powerful that god is. If my very limited capacity for compassion and love transcends that of god's, then that is a small god. Might dose not make right. But I believe in the god who calls us out of the boat (our cosmological arks and paradigms) and onto the waters of chaos themselves. The god who fathers calamity, who the lightning bolt report to and who both the storm clouds and proud waves obey. I understand the need for a firm foundation but I'm far more concerned with being breathed of dust. With having a spirit of love and not fear. I chose compassion, and love over fear, shame and hatred any day. I chose heaven on earth even under threat of hell. I chose true life even at the cost of the death of a lie. And for this I will not apologize! I believe in a god who loves their creation as they made it and as it is. And I will not settle for a small egotistical, dogmatic, homophobic racist, sexist, or nationalistic god. I chose to be human and join the heartbroken in their grief and sorrow because it is a beautiful expression of life (John chapter 11:35). I chose to weep with the god who sees our fickle, fleeting and fragile humanity as precious and beautiful. Because our proclivity for corruption and decay is a sign of our capacity for adaptation and growth. There is profound wisdom and even prolific messages in these ancient traditions but if we don't let them grow and evolve they will stagnate. Sterility is not sanctity! If we cling to this once good fruit it will rot and become bitter in our mouths. The abundance of life is not in the order that tries to preserve it but in the chaos that tests it and delivers it from atrophy. This is what the whole story has been about from the very beginning. Before god became a product sold to solve humanity, because our humanity is not derogatory. But if it is, "I'd rather be whole then good"

Dear reader. I am aware that this book was chaos incarnate, it was by no means clean or concise but nether are we. The point of this literation was to articulate that it is all interconnected and integral. From the tangents to the rabbit holes, they are all the point. The ineffable is a word used to describe god. But more precisely refers to that which transcends empiricism, that which cannot be quantified.

There is a common fear, that you may wake up one day and you won't even know who you are anymore. Now seeing as how it is the primary prerogative of our genes to procreate likeminded genes. it is no surprise then that those genes which exhibit a proclivity for mutability would then be judged corrupt and even evil. Or that our proclivity for corruption and decay then equates to humanity being inherently sinful or corrupt. However the alternative is sterility and that most certainly does not equate to sanctity. Tho it may be devoid of flaws, it is also absent of life. What this degradation of humanity as a broken thing that needs to be fixed, fails to realize is that it is that same proclivity for corruption and decay which then give us the capacity for adaptation and growth. It was the fall of man, when we became conscious and capable of altering those prerogatives incentivized by our instincts which then allowed us to learn and even cultivate compassion for those traits which display a profound dichotomy from ours. It is not the fleeting and fragile nature of humanity nor is it the compassionate and intuitively curious persona which is the problem to be solved. But rather the systems of power which have cultivated a false sense of scarcity in the form of dualistic identities. the problem isn't that we change but rather that somewhere along the way we become comfortable with an idea and stop rediscovering what it means to be us! Because far too often we become more concerned with retaining the security and stability of a stationary identity then we did with developing the capacity for new discoveries. The beautiful truth is you will never know yourself fully, but will instead spend your entire life rediscovering what it means to be you, here and now, in all of life's shapes and forms.

Last minute quotes

Of corse all the hippies have straight teeth. Conservative habits stagnate without progressive mentalities. Preservation without progress pasteurizes any potential for prosperity to propagate.

There will always be those who crave chaos. We were never meant to settle, even for perfection itself. After all How boring would that be! This perpetual hunger is not simply innate, but intrinsic to humanity in the way it inspires change and inhibits the inert. This publication does not pursue perfection but prosperity.

Oscar wild once claimed "the only excuse for a useless thing such as art, is that it is adored immensely" and perhaps the same it true for the futility of life.

The only value innate to life is in one's ability to appreciate it as valuable! (Cancer is a living organism and yet it is nether conscious or compassionate.)

Love is not about solving the problem of gravity but recognizing the ground your tethered to is holy. Love cannot be found only realized. Even so there will always be things we have not the eyes to see. The intuitive heart knows words the rational mind does not speak, each see beauty where other sees only beast.

"If you know exactly what you want to be in life, you will become it. That is your punishment." (Oscar Wilde)

The end

ABOUT THE AUTHOR

My name is Matthew Ryan Holt. I am not an prestigious scholar or renowned influencer. I am an electrician by trade and I hold no high degree in any scholarly field and further more I was diagnosed with dyslexia around 5th grade. The truth of the mater is I am for all intensive purposes a no body. And I speak by no other authority other than that as a human being of approximately 30 years of age.

I grew up as a fundamentalist evangelical Christian and have recently become an agnostic as an attempt to relinquish my religious And political identities. I am an enneagram 4, an INFJ in myers brigs and an avid photographer (all of the photos and art in this book are of my own creation). I am all of these things consecutively and yet none of them In exclusivity.

ACKNOWLEDGEMENTS

I have been privileged to bare witness to a great number of great minds who have inspired this work. From teachers such as rob bell to c.s.Lewis, from rabbi Abraham heschel to Jordan Peterson. This list includes psychologists such as Steven pinker, Carl Jung and Hillary bicbride. From philosophers like Fredrick neiztchie and story tellers like L.M.Montgomery. to theologians like dr Tim mackie and dr Dan McClellan. And Tho my father and mother would have strongly disagreed with my conclusions in this book. I was still blessed with loving family both now and as I was growing up.

Printed in the United States
by Baker & Taylor Publisher Services